LUMINAIRE
光启

守望思想　逐光启航

# 建筑可阅读

## 这里是上海

STORIES OF SHANGHAI ARCHITECTURE

宗明 主编

上海人民出版社 LUMINAIRE BOOKS 光启书局

上海城市推广中心　出品

Produced by Shanghai City Promotion Office

# 出版前言

寻找城市的魅力，建筑是一个绕不过去的窗口。人们常常把建筑视作印刻城市文化特征与时代风格的一个载体、一个见证。什么样的城市，孕育了什么样的建筑。同样，什么样的建筑，也塑造了什么样的城市。某种意义上，建筑就是城市本身。阅读一座建筑、阅读一片街区、阅读一段城市建筑发展史，就是在阅读这座城市。

2019 年 11 月，习近平总书记考察上海，在黄浦江畔提出了“人民城市人民建，人民城市为人民”的重要理念，深刻揭示了中国特色社会主义城市的人民性，赋予了上海建设新时代人民城市的新使命，为我们以更高的政治站位、思想起点谋划推进城市工作，更好地体现时代性、把握规律性、富于创造性，指明了前进方向、提供了根本遵循。上海市委书记李强在市委第十一届九次全会上提出，要打造人人都有人生出彩机会、人人都能有序参与治理、人人都能享有品质生活、人人都能切实感受温度、人人都能拥有归属认同的城市。要以软实力提升彰显人民城市精神品格，大力弘扬上海城市精神和城市品格，不断彰显红色文化、海派文化、江南文化的独特魅力，更好地延续城市文脉、保留城市记忆。让人在城市也能“诗意地栖居”。

历史建筑承载历史记忆，彰显城市底蕴。上海，素来有“万国建筑博览会”之称，汇聚了不同时期、不同风格的建筑，讲述着不同的故事，处处积淀着“海纳百川、追求卓越、开明睿智、大气谦和”的城市精神，处处彰显着“开放、创新、包容”的城市品格，处处承载着红色文化、海派文化、江南文化交相辉映的文化特质。李强书记指出，上海丰富的旅游文化资源，完全可以满足“听故事、品生活，找个地方静静发呆”。上海要学会讲故事，特别是中心城区的历史建筑、历史街区，故事要讲活，要让人“可阅读”。我们阅读梧桐树下老洋房、老弄堂里石库门，阅读一处处名人故里、历史遗存，阅读从昔日“工业锈带”变身的今日“生活秀带”，阅读数百年甚至两千年前就留下的亭台楼阁桥寺塔，阅读被视作改革开放经典象征的东方明珠、陆家嘴“三件套”，阅读嵌在人们日常生活中的文化地标……我们阅读的是一个个独具美感的建筑空间，更是背后一座城市波澜壮阔的发展史、进步史、创新史。不同的建筑从不同侧面，提示着上海这座城市所特有的地位与身位——这是中国共产党的诞生地，是中国共产党人初心孕育、梦想启航的地方，红色基因、红色血脉，百年传承不息；这是一座具有鲜明社会主义性质、富于现代化气息的国际大都市，是多元文化荟萃之地，是中国拥抱世界的前沿，也是世界观察中国的窗口；这是一座对内对外开放两个扇面的枢纽城市，是国内大循环的中心节点、国内国际双循环的战略链接；这是一座地处江南中心地带、承载江南文化衣钵的城市，粉墙黛瓦、小桥流水、枕水而居的千年传统，至今还留有鲜明的印记，也成为城市的底色；这更是一座时时领风气之先、始终奋进新时代、立志创造新奇迹的城市，从不断刷新的上海高度、上海速度里，能读到中国的“高度”、中国的“速度”、中国的“温度”；这终究是一座人民的城市，传奇般的建筑空间终究来自人民的智慧与双手，它们也终究是属于人民的财富与记忆。

“建筑可阅读”，建筑本身是有故事的，要把故事留下来，自然要把建筑的肌理保护好；建筑身上的故事不应该“养在深闺人未识”，理应是被尽可能多的人读到、听到、看到、体验到的。这些年，上海在推进城市更新和精细化管理的过程中，投入大量资金、资源、人力，开展了一系列优秀历史建筑保护活化工程，在空间上“修旧如旧”，在功能上活化利用，把故事讲起来，让建筑活起来，并立足整体风貌保护，对标最高标准、最好水平，坚持长远眼光、延续城市文脉。城市更新，不仅要尽最大可能保留保存城市风貌和历史建筑，使之可阅读、可展示，也要充分考虑民生功能、公共服务的留存拓展，让空间更有温度、更富活力、更好彰显城市魅力。因此，大力推进历史建筑对外开放，打造“街区漫步”等城市微旅游产品，并致力塑造更多可亲近、能共享的公共空间，在历史文脉保护中诠释人民立场。同时，把“可阅读”的思路贯穿到新建建筑的设计、建设之中，不断提升城市规划与设计的品质，致力打造更多的城市精品。要让五湖四海的人向往这座城市、汇聚到这座城市，“在成就自己梦想的同时，造就一座伟大的城市”。

建筑的故事，背后也是上海故事、中国故事。记录故事、讲述故事、传播故事，需要城市治理方方面面的主体共同努力，需要付出热心与情怀。现在这部《这里是上海：建筑可阅读》，就是努力的一部分。

2019 年 12 月，根据李强书记关于加强“建筑可阅读”文旅融合工作的指示精神，上海市副市长宗明主持启动《这里是上海：建筑可阅读》一书的编辑出版工作并担任主编，上海市文化和旅游局（市文物局）与上海世纪出版集团、上海人民出版社、上海城市推广中心共同组织、联手打造，上海建筑规划、文物保护、文学艺术、新闻出版界等领域一批著名学者专家共同参与。全书遴选了 56 处能充分体现上海城市品格，体现红色文化、海派文化、江南文化和改革开放崭新形象的代表性建筑。这些建筑跨越不同时代，兼顾了城市建筑与其他建筑、西式建筑与中式建筑、外国设计师作品与本土设计师作品等不同的类型风格，包含了全市范围内重要的全国重点文物保护单位、上海市文物保护单位和上海市优秀历史建筑，以及一些独具特色的建筑。在本书编辑出版过程中，各建筑保护与管理单位积极支持，多位长期从事城市历史和建筑文化研究的专家学者以及长期从事城市影像记录的摄影师热情投入，就内容文字和图片素材反复打磨，为图文并茂地呈现“建筑可阅读”提供了质量保证。

“世界那么大，是不是先到上海来看看?”希望读者能透过这部书，打开上海城市建筑这部“大书”，走进上海这座“人民城市”，在这里阅读上海、读懂上海，在这里体验上海、爱上上海。

2020 年 7 月

# Preface

To discover the charm of a city, architecture can be a window that no one shall miss. People often regard architecture as a showcase for and a witness to a city's cultural characteristics and styles of different times. A city breeds its architecture. Likewise, architecture also shapes the city. In a sense, architecture is the city itself. When you look at a building, survey a block, or browse a piece of architecture history in a city, you are actually reading the city.

In November 2019, during his inspection tour in Shanghai, General Secretary Xi Jinping proposed the important concept that "a people's city is built by the people and for the people", which underscores a people-centred development philosophy of socialist cities with Chinese characteristics and gives Shanghai a new mission – to build a people's city in the new era. It also points the way and provides a fundamental guideline for us to proceed with a greater purpose and from a higher starting point in planning and advancing city operations, so as to better embody the spirit of the times, reflect a mastery of the laws of development, and demonstrate creativity. At the 9th Plenary Session of the 11th CPC Shanghai Municipal Committee, Li Qiang, Secretary of the CPC Shanghai Municipal Committee, proposed to create a city in which everyone has a chance to shine in life, participates in governance, enjoys a quality life, feels love and warmth, and has a sense of belonging. We should build the character of the people's city with soft power, promote the ethos and character of Shanghai, and highlight the unique charm of the revolutionary traditions of the CPC, Shanghai-style culture, and the Jiangnan culture so as to better preserve the heritage and memories of the city and enable people to "dwell poetically" in the city.

Historical buildings bear the memories of history and reveal the profound culture of a city. Shanghai, well-known for its "Exotic Building Clusters", is home to buildings of different periods and architectural styles that tell varied stories to the world. The character of openness, innovation and inclusiveness can be found in every respect of the city, along with the convergence of revolutionary traditions, Shanghai-style culture and the Jiangnan culture. As Secretary Li Qiang pointed out, Shanghai is rich in tourism and cultural resources and can fully meet people's need to "listen to stories, get a better taste of life, or find a place to spend the peaceful leisure time". Shanghai needs to learn how to tell stories, especially about the historical buildings and historic blocks in the downtown area. The stories need to be told creatively and make the buildings "readable". When we read the old Western-style houses under the phoenix trees, the Shikumen in old lanes and alleys, the former residences and legacies of the noted figures, today's "greenbelt" transformed from the past "rustbelt", the pavilions, terraces, open halls, bridges, temples and pagodas from several hundred years or even two thousand years ago, the reform and opening-up icon Shanghai Oriental Pearl Radio & TV Tower and the trio of signature skyscrapers at the heart of Lujiazui, and the cultural landmarks embedded in people's daily lives, etc., we are reading not only architectural spaces with unique beauty, but also the city's magnificent history of development, progress and innovation. Different buildings tell about the special status of Shanghai from different angles. This is the birthplace of the Communist Party of China (CPC), where the CPC members conceived their original aspirations and set sail to pursue dreams. The revolutionary traditions have been passed down for a hundred years. This is an international metropolis rich in modernity with distinctive socialist characteristics. Here diverse cultures converge while China's readiness to embrace the world is best manifested, and foreigners can get a closer look at China. This is a hub city opening to and connecting both domestic and international markets, a central node of the economic network across the country and a strategic link between domestic and global connections. This is a city located in the centre of Jiangnan where cultural tradition featuring the distinctive marks of white walls and black tiles and streams flowing beneath bridges has been well preserved. This is a city that always leads the trend, strives to advance in the new era, and is determined to create new miracles. The record height and speed Shanghai has reached is also telling about China's drive to scale new heights in the pursuit of better development. After all, this is a people's city. The legendary architectural spaces are the outcomes of people's wisdom and labour and therefore part of people's wealth and memory.

Architecture is readable. The architecture itself has stories. To pass the stories down, it is necessary to protect the architecture. Such stories should not be concealed, but should be read, heard, seen and experienced by as many people as possible. In recent years, in the process of promoting urban renewal and lean management, Shanghai has invested substantial funds, resources and manpower in historic preservation projects, restoring the original appearance of the old buildings and reviving them with new functions. Ultimately, the purpose is to get the stories told, bring the buildings back to life, conserve the overall landscape in line with the highest standards, and preserve the city's cultural traditions for the future. Urban renewal is not only about trying to conserve and retain the cityscape and historical buildings as much as possible so that they can be read and displayed, but also about giving full consideration to the preservation and expansion of functions and public services concerning people's lives so that the spaces give people a greater sense of warmth and vitality and better showcase the charm of the city. Hence the need to open historical buildings to the public, build tourist blocks in the neighbourhoods and create more accessible public spaces. Moreover, the idea of "readability" should be applied to the design and construction of new buildings to improve the quality of urban planning and to create more quality spaces for the city. The purpose is to draw people from all over the world to this city so that they can help to make the city greater while fulfilling their dreams.

The stories behind architecture are also the stories of Shanghai and of China. To record, tell and spread the stories require joint efforts as well as passion and devotion of all parties involved in the city governance. The book *Stories of Shanghai Architecture* is part of the effort.

In December 2019, according to Secretary Li Qiang's instruction on strengthening culture and tourism integration with stories of architecture, Zong Ming, Deputy Mayor of Shanghai, presided over the editing and publishing of the book *Stories of Shanghai Architecture* and acted as the editor-in-chief. The endeavour is sponsored by Shanghai Municipal Administration of Culture and Tourism (Shanghai Municipal Administration of Cultural Heritage), Shanghai Century Publishing Group, Shanghai People's Publishing House, and Shanghai Urban Promotion Centre, involving famous scholars and experts in Shanghai's architectural planning, cultural heritage protection, literature and art, press and publishing circles and other fields. The book covers 56 iconic buildings that can fully reflect the character of Shanghai, the revolutionary traditions of the CPC, Shanghai-style culture, the Jiangnan culture and the new image of reform and opening up. These buildings span different periods of history and encompass various types and styles such as urban and other buildings, Western-style and Chinese-style architecture, the works of foreign and Chinese architects, as well as the Major Historical and Cultural Sites Protected at the National Level in the city, Monuments under the Protection of Shanghai Municipality, Heritage Architecture of Shanghai Municipality as well as some buildings with unique features. In the process of editing this book, various architecture conservation and management organizations rendered active support and a number of experts and scholars who have long been engaged in the study of urban history and architectural culture, as well as photographers who have been engaged in urban image recording devoted themselves to the project and helped to polish the text and image materials repeatedly. Their contributions have guaranteed the quality of the book.

The world is so big, and why not come to Shanghai to see more of it? This book is intended to offer readers a graphic account of Shanghai's architecture and an intriguing glimpse of Shanghai as a people's city so that they can better understand Shanghai, come to experience it and fall in love with it.

July 2020

# 目录
# Contents

## 都会映象 Metropolis Impression

## 梧桐深处 Amidst the Plane Trees

# 春申古风

Ancient Heritage

松江唐经幢
Songjiang Sutra Pillar of the Tang Dynasty

松江唐经幢全名为“佛顶尊胜陀罗尼经幢”，建于唐大中十三年（859 年），现坐落在松江区中山小学内，是上海地区现存最古老的地面文物，也是全国唐代经幢中最完整和高大的一座，1988 年被公布为全国重点文物保护单位。

这是当时松江笃信佛教者蒋复、沈直轸为超度父母以及其他家人而造。至 20 世纪 60 年代初，该经幢仅有十级露出地面，其余均埋于土内。经发掘、修复始恢复全貌。经幢上记载的文字，将遥远的唐代社会风俗一下子真切地还原到眼前。

经幢为石灰岩材质，现存 21 级，高 9.3 米。幢身分上下两截，均为八角形，直径达 76 厘米。上段刻有《佛顶尊胜陀罗尼经》全文，下段刻有题记及捐助者姓名；幢身下设勾栏平座、叠涩，其下有三重束腰、三层托座，从下至上三段束腰上分布刻盘龙、蹲狮及菩萨壶门，三层托座分别为海水纹底座、莲瓣卷云座、唐草纹莲座。幢身上部共有十级，从下至上有狮首华盖、联珠、卷云纹托座、四天王座像、八角腰檐、仰莲托座、礼佛浮雕圆柱、八角攒尖盖等。经幢通体雄伟秀丽、造型优美、层次组合匀称，幢身浮雕生动形象，技法洗炼圆熟，极具盛唐艺术风格。

松江区中山东路43号中山小学内
Inside Zhongshan Primary School,
43 East Zhongshan Road, Songjiang District

The Usnisa Vijaya Dharani Sutra Pillar in Songjiang was built in 859, the 13th year of Dazhong Period of the Tang Dynasty. It now stands in Zhongshan Primary School of Songjiang District as the oldest aboveground cultural relic in Shanghai, and also the most complete and tallest sutra pillar of the Tang Dynasty in China. It was announced as a Major Historical and Cultural Site Protected at the National Level in 1988.

The Pillar was built by the pious Buddhist believers Jiang Fu and Shen Zhizhen at that time to release souls of their deceased parents and other family members. Up to the early 1960s, only ten levels of the building were exposed, and the rest were buried underground. After excavation inscriptions, the whole pillar was restored to full view. These inscriptions bring back the social customs of the Tang Dynasty before our eyes all of a sudden.

The pillar is made of limestone, 9.3 metres high, and with 21 levels remained. It is divided into upper and lower sections, both in octagonal shape, with a diameter of 76 centimetres. The upper section engraved with the full text of Usnisa Vijaya Dharani Sutra, and the lower section engraved with inscriptions and names of donors. The lower part of the pillar was built with a flat seat, corbellings, and three girth piers and three-layer pedestals beneath, which are carved with coiling dragons, crouching lions and Bodhisattva Kunmen, as well as ripple patterns, lotus-petal designs, and scroll designs respectively. There are ten levels in the upper part of the pillar, including the lion-head canopy, the granulation, the cloud patterned bracket, the sitting statues of the four Heavenly Guardians, the octagonal waist eaves, the lotus bracket, the Buddha worshipping relief cylinder, and the octagonal conical roof from bottom up. The whole structure of the pillar is magnificent and beautiful with graceful shape and well-balanced level combination. The relief on the pillar is vivid and sophisticated in workmanship, rich in the artistic style of the Golden Age of the Tang Dynasty.

松江方塔
Songjiang Square Pagoda

兴圣教寺塔位于美丽的松江方塔园内，因其平面为方形，人们习惯称其为“方塔”。兴圣教寺塔除了第七至九层系清代重建，其余各层的构件大部为宋代原物，不少地方保存了唐、五代的木作手法，尤其是斗栱保留了宋代原物的 60% 以上，是江南古塔中保留原构件较多的一座。1996 年被公布为全国重点文物保护单位。

兴圣教寺塔塔身修长，共有九层，高达 42.5 米，是上海现存次高的古塔。该塔虽建于北宋，但因袭唐代砖塔形制，呈四方形，具有唐代建筑风格。塔底层边长为六米，往上逐层收缩。砖塔外壁为“砖夹木”，每隔五六皮至十余皮砖会嵌入一根长方形的横木，以加强墙体。塔身每面被砖柱分为三间，正中设壶门。每层设木质平座、栏杆，有木斗栱承托。底层楼梯在围廊内，二层以上楼梯均在塔身内。顶层为攒尖式屋顶，长达 13 米的塔心木在八层设支撑，穿出屋面八米多，套以铁质塔刹，并有覆盆、露盘、相轮、宝瓶等。有四根铁索从塔尖拖向顶层的檐角，塔檐四角设铜铃，以驱赶飞鸟落塔做窝。

该塔在南宋和元明时曾多次修葺，清乾隆年间又经历大修。在 1974—1977 年重修过程中，新发现不少宋代文物，保存了部分宋代原构件。1978 年，上海市园林局以方塔为中心建成方塔园。

松江区中山东路235号方塔园内
Inside Songjiang Fang Ta Park,
235 East Zhongshan Road, Songjiang District

Located in the beautiful Square Pagoda Garden, the Square Pagoda of Songjiang got is name for the building is square. Except the storeys from the 7th to the 9th, which were reconstructed in the Qing Dynasty, the pagoda basically retained its original components in the Song Dynasty, especially the bracket system, over 60% of which are the Song objects. A great part of the pagoda was built with the woodwork techniques of the Tang and Five Dynasties. It is one of the ancient pagodas that retain the most original components in southern China. In 1996, it was announced as a Major Historical and Cultural Site Protected at the National Level.

The 9-storey Square Pagoda looks tall and slender, about 42.5 metres high. It is the second highest existing pagoda in Shanghai. The pagoda was built in the Northern Song Dynasty, but largely inherited the shape and structure of the brick pagoda and architectural style of the Tang Dynasty. The side length of the ground floor is 6 metres and the length decreases storey by storey upward. The outer wall of the brick pagoda is "brick with wood", and a rectangular cross beam is embedded every five or six to more than ten brick layers to strengthen the wall. Each side of the pagoda is divided into three bays by brick colums, with the Kunmen set in the centre. Each floor is built with wooden flat seats and railings, supported by wooden bracket system. The stairs on the ground floor are built in the arcade, and the stairs from the second floor and up are in the body of the pagoda. The top floor is the conical roof. The 13-metre wooden pagoda core column is supported at the eighth floor and sticks more than 8 metres above the roof. It is covered with iron pagoda spine, and installed with statue seat, amrta-kalasa and other Buddhist components. There are four iron cables dragging from the spire to the eaves corner of the top floor. Copper bells are set at the four corners of the eaves, to keep the birds from making nests on the pagoda.

The pagoda went through many times of betterment in the Southern Song, Yuan and Ming Dynasties, and was heavily repaired in the reign of Emperor Qianlong of the Qing Dynasty. In the process of rebuilding from 1974 to 1977, a lot of relics of the Song Dynasty were discovered and some original Song components were kept. In 1978, the Shanghai Landscape Bureau built the Songjiang Fang Ta Park with the Square Pagoda as its main building.

# 龙华塔和龙华寺

Longhua Pagoda and Longhua Temple

徐汇区龙华路2853号
2853 Longhua Road, Xuhui District

上海这座国际大都会，百年来向以高楼大厦著称，假设我们穿越回公元 10 世纪，去寻找这片土地上的最高建筑，大概非龙华寺内的龙华塔莫属。

龙华寺的历史渊源颇为传奇。据清人张宸记载，三国吴赤乌五年（242 年），来自西域的康僧会在从交趾前往建业（今南京）的路上过龙华荡，见“水天一色，藻荇交横”，是“尘辙不到，颇宜清修”之地，遂在此“建立茅茨，设像行道”。后康僧会到建业觐见了吴主孙权，又获赐修建 13 座舍利佛塔。

此后一千多年来，龙华寺几度重建，历经衰荣。现存的龙华寺建筑群主体建于清同治、光绪年间，保存了宋代禅宗的“伽蓝七堂制”。中轴线上有牌坊、山门，第一进为弥勒殿，两侧为三重飞檐的钟楼、鼓楼，第二进为天王殿，第三进是面阔五间的大雄宝殿，第四进为三圣殿，第五进为方丈室，第六进为藏经楼，其中大雄宝殿、三圣殿为重檐歇山顶，气势恢弘。1959 年，龙华寺被列为上海市文物保护单位。

龙华塔与龙华寺相对而立，其塔身、塔基为北宋所建原物，距今已有一千余年历史。20 世纪 20 年代，其平座、栏杆曾被施以钢筋水泥，1954 年，经历重修，恢复了原貌。

龙华塔为楼阁式砖木结构，七级八面，塔高 40.347 米。塔身外壁为八角形，砖砌角柱。塔身内为方室，底层高大，向上各层面积与高度逐渐收缩，形成密檐。每层四面设壶门，四面有壁龛，且逐层转换，使外立面上各层门窗依次变换。各层楼板下隐出砖栱，栱头卷刹分三瓣，外檐转角铺作鸳鸯交首栱，底层围廊柱子的柱头呈梭状，枋上有七朱八白之装饰。龙华塔的基础采用木桩加木承台的方式，体现了工匠在软土地基上建高层建筑的智慧。2006 年，龙华塔被公布为全国重点文物保护单位。

Shanghai, as an international metropolis, has been famous for its high-rise buildings for over a century. If we were to go back to the 10th century to look for the tallest building in the city, it probably would be the Longhua Pagoda in Longhua Temple.

The historical origin of Longhua Temple is quite legendary. As recorded by Zhang Chen of the Qing Dynasty, in the 5th year of the Chiwu Period of the State of Wu (242) in the Three Kingdoms period, Kang Senghui, a man from the West Regions, passed by Longhua Lake on his way from Jiaozhi to Jianye (present-day Nanjing). He found that the water and the sky were merged into one color, with green weeds everywhere; a place far from the crowds and peaceful for meditation. So he set hands to build a temple there. Later, Kang Senghui went to Jianye and presented himself to Sun Quan, the ruler of the State of Wu, and was granted permission to build 13 stupas.

Since then, for over a thousand years, Longhua Temple has been rebuilt several times and has gone through ups and downs. The main body of the existing Longhua Temple complex was built in the Tongzhi and Guangxu periods of the Qing Dynasty, retaining the "seven-halled temple" type of Zen in the Song Dynasty. There are memorial archways and mountain gates on the central axis. The first row is the Maitreya Hall, with the Bell Tower and Drum Tower in triple-layer cornices on both sides. The second row is the Heavenly King Hall. The third row is the five-room wide Great Shrine Hall. The fourth row is the Hall of the Three Sages. The fifth row is the Abbot's Room, and the sixth row is the Scripture Tower. Among them, the Great Shrine Hall and the Hall of the Three Sages are in the styles of double eaves and East Asian hip-and-gable roof, very magnificent. In 1959, Longhua Temple was listed as a Historical and Cultural Site Protected at the Shanghai Municipal Level.

The Longhua Pagoda stands opposite Longhua Temple. Its body

and base were originally built in the Northern Song Dynasty, boasting more than a thousand years of history. In the 1920s, its flat seat and railings were reinforced with cement, and in 1954, it underwent a renovation and was restored to the appearance of the pagoda.

Standing 40.347 metres tall, the Longhua Pagoda is a brick and wood structure with seven storeys and eight sides. The outer wall of the pagoda is octagonal and is decorated with red wood-like brick beams. The interior is a square space, with the bottom floor big and tall, and the area and height of each floor gradually shrinking upward to form dense eaves. Each storey, with eight sides, is set with Kunmen (like a doorway) style doors and shrines in an alternating manner, and the positions of these doors and shrines rotate by 45 degrees on each floor. Under each floor is a hidden brick arch. The entasis of each arch is divided into three parts. The corner on the outer eaves is paved in an arch style named "yuanyangjiaoshou". The column heads of the colonnaded walkway around the pagoda on the ground floor are spindle shaped, and the square-column is decorated with seven red decorative features and eight white ones. The foundation of the Longhua Pagoda adopts the way of timber pile and pile cap, which reflects the wisdom of artisans in building high-rises on soft soil foundation. In 2006, the Longhua Pagoda was listed as a Major Historical and Cultural Site Protected at the National Level.

# 嘉定孔庙

Jiading Confucian Temple

嘉定区嘉定镇南大街183号
183 South Street, Jiading Town, Jiading District

嘉定孔庙（又名文庙、庙学或学宫）由嘉定县第一任知县高衎孙始建于南宋嘉定十二年（1219 年），建筑群占地 1.8 公顷，建筑面积 1.1 万平方米，规格形制保存较为完好，是上海地区现存规格形制保存最为完好的孔庙。2013 年被公布为全国重点文物保护单位。

现存的建筑群大致分为东、西两片。西片以原孔庙大成殿为中心，南北成一条中轴线，依次有牌坊、棂星门、泮池和三桥、大成门、大成殿；东片以原县学明伦堂为主体，两旁设有东西庑（后有碑廊），前有礼门三间，棂星门外有三座精雕细刻的石牌坊（兴贤坊、育才坊、仰高坊），沿汇龙潭的石栏上雕有形态各异的 72 尊石狮，象征孔门七十二贤。大成殿建立于石台之上，面阔与进深均为五间，其中殿身为三间，其余各间为回廊，为双檐歇山顶，檩、枋上有木纹彩绘。明伦堂面阔五间、进深三间，前出抱厦三间。整个平面为“凸”字形。大成殿、明伦堂的梁架结构比较特别，它们都采用了横跨六步架的大月梁，省去了门前金柱两根，扩大了内部空间，具明代风格。

清雍正初年，明伦堂东侧建兴文书院，后于乾隆二十二年（1757 年）改为应奎书院。清乾隆三十年（1765 年），书院又因陆陇其号，易名为当湖书院。当湖书院是上海地区仅存的清代书院。

嘉定孔庙的不断修建还促成了应奎山、汇龙潭的产生：明天顺、正德、万历年间，当地士人为破解开门见寺、不利孔庙的风水格局，决定在庙前挖土筑山，并取名为“应奎山”，以遮蔽庙南的留光寺。因挖土堆山，形成了较大的水面，遂引水环绕，疏浚了应奎山周围的五条河流，使之呈“五龙抱珠”之势，遂成“汇龙潭”。

如今，嘉定孔庙设有上海中国科举博物馆，相关内容陈列能让游客直观地领略古代考试制度与科举文化的发展历程。

Jiading Confucian Temple was built in 1219, the 12th year of Jiading Period of the Southern Song Dynasty, by Gao Yansun, the first magistrate of Jiading County. The building complex covers a land area of 1.8 hectares and a floor area of 11,000 square meters, with relatively intact shape and structure. It is the most well-preserved Confucian Temple in Shanghai. In 2013, it was announced as a Major Historical and Cultural Site Protected at the National Level.

The existing building complex is divided into east and west sections. The west section takes the Hall of Achievements of the original Confucius Temple as the centre, and on its central axis from south to north are the memorial archway, Lingxing Gate, Panchi pond and three bridges, and the Gate of Achievements and the Hall of Achievements. The east section takes the former county school Minglun Hall as its main building, with east and west side rooms (with Stele Gallery behind) and three ceremonial archways in the front. There are three finely carved stone archways (Xingxian Archway, Yucai Archway, Yanggao Archway) in front of the Lingxing Gate. The stone railings along Huilong Pond are carved with 72 stone lions in different shapes, symbolizing the 72 disciples of Confucius. The Hall of Achievements is built on a stone platform, with five bays in width and depth. The hall takes up three bays, and the rest bays are ambulatories. It has a double hip and gable roof structure. The purlins and columns are painted with wood grain in different colours. The Minglun Hall is five bays wide and three bays deep, and have porticoes on three sides. The entire plane is like an upturned "T". The beam frame structure of the Hall of Achievements and Minglun Hall is quite special that the big crescent beams span a six-step frame, the two principal columns in front of gate are removed, and the internal space is thus expanded, which is typical of the Ming Dynasty style.

In the early years of Emperor Yongzheng's reign in the Qing

Dynasty, Xingwen Academy was built in the east of Minglun Hall, and it was renamed Yingkui Academy in 1757, the 13th year of Emperor Qianlong's reign. In 1765, the 30th year of Emperor Qianlong's reign, it was renamed Danghu Academy after the literary name of the honest official Lu Long. The Danghu Academy is the only existing academy of the Qing Dynasty in Shanghai.

The continuous construction of Jiading Confucian Temple also contributed to the emergence of Yingkui Hill and Huilong Pond. In the Tianshun, Zhengde and Wanli periods of the Ming Dynasty, local scholars decided to build a hill in front of the temple to safeguard the temple for matters of fengshui, and named it "Yingkui Hill" to shade the Liuguang Buddist Temple on the south of the confucion temple. As a result of digging and piling up a hill, a large water surface was formed, and water was diverted and five rivers around Yingkui Hill were dredged to make them look like "Five Dragons holding a pearl", which gave the pond the name of "Huilong Pond (Dragon Gathering Pond)".

Today, there is a Chinese Imperial Examination Museum in Jiading Confucian Temple. The exhibits in the museum are intended to help visitors to better understand the history of China's ancient examination system and imperial examination culture.

普济桥

Puji Bridge

青浦区金泽镇
Jinze Town, Qingpu District

青浦区的金泽古镇位于京杭大运河东侧，始建于南宋时期。这里有“江南第一桥乡”之美誉，现仍存宋元时期古桥数座，其中普济桥是上海地区建造年代最早、保存最完好的石拱桥。现为上海市文物保护单位。

普济桥建于南宋咸淳三年（1267 年），为单孔石拱桥，因桥畔有圣堂庙，故也称“圣堂桥”。建桥所用石料为珍贵的紫石，每当雨过天晴，阳光照射桥上，桥体会散发光泽，晶莹如宝石，因此人称“紫石桥”。

普济桥全长 26.7 米，宽 2.75 米，跨径 10.5 米，采用椭圆券形的弧拱，结构上比半圆拱省工省料，且拱跨较大，坡度平缓，展现了高超的建造水平。该桥的拱券砌筑形式与著名的赵州桥相同，采用拱券石并列砌置的方法，但两排并列拱板之间略有错开，单双不一。两堍有引桥，水平连接石端部券石上刻有莲幡状图案，桥栏紧贴桥面形成柔和的曲线，远眺如月牙，纤巧飘逸，非常迷人。

该桥在明清两代经过重修，清雍正初年原样重整了石栏杆。桥顶处有木框架子，装置晚间可关闭的木门。现该桥桥体夹杂有青石、花岗石等。

Jinze Ancient Town in Qingpu District, located on the east side of the Grand Canal, was finest founded in the Southern Song Dynasty. Jinze is hailed "the first bridge town in Jiangnan". Several of the ancient bridges from the Song and Yuan Dynasties still exist. Among them, the Puji Bridge is the earliest-built and best preserved stone arch bridge in Shanghai. Now it is a cultural site protected by the city of Shanghai.

Built in 1267, the Puji Bridge is a singlespan stone arch bridge and beside it is the Shengtang Temple. So, the bridge is also called "Shengtang Bridge". The bridge was built with precious purple stones. When the sun shines on the bridge after rain, the bridge will emit lustre, glittering and translucent like gems, so it is also known as the "Purple Stone Bridge".

The Puji Bridge is 26.7 metres long, 2.75 metres wide and with a span of 10.5 metres. It adopts the elliptic arch, which costs less labour and material than the semicircular arch in structure. Moreover, the arch span is larger and the slope is gentle, showing the superb construction skill. The arch masonry form of the bridge is the same as that of the famous Zhaozhou Bridge, with the arch stones placed side by side. But the two rows of parallel arch bars are slightly staggered in a disordered fashion. There is an approach on either end of the bridge. There are intertwined lotus patterns carved on the arch stone connecting the ends levelly. The bridge railing is close to the bridge deck to form a soft curve. It looks like a crescent moon in the distance, graceful and fascinating.

The bridge was rebuilt in the Ming and Qing Dynasties, and the stone railings were reconstructed in the early Yongzheng Period of the Qing Dynasty. There are wood frames on the top of the bridge to install a wood door that can be closed at night. Now the bridge body is mixed with bluestones and granites.

**真如寺大殿**

Mahavira Hall of Zhenru Temple

普陀区兰溪路399号
399 Lanxi Road, Putuo District

我国的传统建筑以木结构居多，保存不易。许多人也许会好奇，上海境内目前最高龄的房屋在哪里？答案是：真如寺大殿。

真如寺大殿位于普陀区兰溪路真如寺内，建于元延祐七年（1320 年），由僧人妙心修建，是上海唯一保存完好的元代佛寺大殿，也是上海现存最古老的木结构大殿，1996 年被公布为全国重点文物保护单位。

真如寺大殿建筑规模并不大，处于现真如寺中轴线上。大殿面阔及进深均为三间，呈方形平面，为接近正方形的九宫格布局，面积达 158 平方米。该建筑原为单檐歇山顶。清末重修时曾被改成五间重檐形式。1963 年，大殿恢复了单檐三间的元代式样。

真如寺大殿内有 16 根梭状柏木柱，其中金柱 6.45 米高，檐柱 4.28 米高。正间的柱身都略向内侧倾斜，金柱内倾 16 厘米，檐柱内倾 8 厘米，其平面布局与著名的元代建筑山西芮城永乐宫龙虎殿相同。斗栱为四铺作小斗栱，后尾斜向挑出。补间铺作当心间置四朵，次间各用二朵。该建筑斗栱粗壮、构架简洁、举折平缓，且屋面硕大，出檐深远，层角起翘平和大方，具元代建筑的典型特点。

Most of the traditional buildings in China are wood structures, and are very difficult to preserve. Many people may ask in wonder, which is the oldest building in Shanghai? The answer is: Mahavira Hall of Zhenru Temple.

Located in the Zhenru Temple on Lanxi Road, Putuo District, the Mahavira Hall of Zhenru Temple was built by Monk Miaoxin in the 7th year of Yanyou Period of Yuan Dynasty (1320). It is the only intact hall of the Yuan Dynasty Buddhist temples of such kind in Shanghai, and the oldest existing wood structure in Shanghai. In 1996, it was listed as a Major Historical and Cultural Site Protected at the National Level.

Not large in scale, the hall is located on the axis of Zhenru Temple. The hall is three bays in width and depth, with a square plane. The layout features a Sudoku pattern somewhat like a square, covering an area of 158 square meters. The building was originally a single-eave structure with a gable and hip roof. Through the renovation in the late Qing Dynasty, it was transformed into a five-bay and double-eave structure. In 1963, the hall was restored to its form in the Yuan Dynasty as a single-eave and three-bay structure.

There are 16 shuttle shaped cypress pillars in the hall. Of which, the principal pillars are 6.45 metres high, and the peripheral pillars are 4.28 metres high. The pillars in the central bay slightly lean inward, with the principal ones leaning by 16 centimetres and the peripheral ones by 8 centimetres. Its plane layout is similar to that of the Dragon-Tiger Hall, Yongle Palace, Ruicheng, Shanxi Province. The bracket system is made up of four small brackets set on columns, with the tails slanting out. There are four brackets resting on the beam in the main bay and two on the beams in the side bays. This building shows typical characteristics of Yuan Dynasty architecture, with the thick and sturdy bracket system, simple structure, low rise of the roof, large roof, deep eaves, and flat and generous angles.

# 豫园
Yuyuan Garden

黄浦区安仁街218号
218 Anren Street, Huangpu District

在上海人的心目中，豫园是一座园林，又远大过一座园林。作为上海老城厢内仅存的古代园林，豫园已有400余年历史，1982年被公布为全国重点文物保护单位。

明嘉靖三十八年（1559年），潘允端初创豫园于上海县城墙内，至万历十八年（1590年）初告建成，由江南名匠张南阳打造，占地70余亩，内有假山水池及30余座亭台楼阁，奇秀甲于东南，可与苏州拙政园、太仓弇山园媲美。明末清初，豫园逐渐破败，部分房屋被用作行业公所，部分园地沦为菜畦。清乾隆二十五年（1760年），一些会馆绅商买下荒废的园子加以整修。1784年，重修后的豫园被当作城隍庙的西园向百姓开放。

20世纪初，豫园被豫园路分为南北两片，南片包括湖心亭、九曲桥、玉玲珑、得月楼、香雪堂等山石楼阁；北片有萃秀堂、点春堂、春风得意楼等。许多建筑被改建成民房，凝晖阁、清芬堂、濠乐舫、绿波廊分别成为菜馆、点心铺、茶楼，形成热闹的庙市、市民的乐园。

1949年后，陈从周主持了豫园的两次大规模整修。1956—1961年，修复了三穗堂、万花楼、点春堂等楼堂，新建了九狮轩、会景楼、玉华堂、曲水流觞等景点。1986—1988年，整修主要集中于东部的玉玲珑、玉华堂、会景楼、九狮轩等景点，重建了环龙桥，修建了积玉山、浣云山、寰中大快照壁和积玉廊，并将原位于沪北天后宫内的古戏台迁入园中。

现在的豫园规模虽仅为明豫园的一半，但其中“大假山”仍为明代原物，三穗堂、点春堂等古建筑形制高大，九曲桥间的湖心亭独具特点，格局仍显江南古典园林之风韵。湖心亭及周围的水池原是豫园初建时的一景，被称为“岛佚亭”“凫水亭”。现在的湖心亭平面接近丁字形，两翼呈多边形，屋顶由六个大小各异的尖锥形和短脊歇山形组成，有 28 只角，错落有致，形态生动。两侧的九曲桥原为石板木栏，1924 年重建时改为水泥栏杆。

For the people of Shanghai, Yuyuan Garden is much more than a garden. As the only existing ancient garden in Shanghai Old Town, Yuyuan Garden can be traced back to more than 400 years ago. In 1982, it was listed as Major Historical and Cultural Site Protected at the National Level.

In the 38th year of Jiajing Period of Ming Dynasty (1559), Pan Yunduan first built Yuyuan Garden within the city wall of the Shanghai County. The construction was completed in early 1590 under the workmanship of Zhang Nanyang, a famous artisan in Jiangnan. Covering an area of more than 70 mu, Yuyuan Garden is reputed for rockeries, ponds and more than 30 pavilions, terraces and open halls, ranking among the top in southeast China and on a par with Humble Administrator's Garden in Suzhou and Yanshan Garden in Taicang. In the late Ming and early Qing, Yuyuan Garden gradually became dilapidated. Some of the houses were used as guild halls, and some of the garden plots turned into vegetable beds. In the 25th year of the Qianlong Period of the Qing Dynasty (1760), the deserted garden was bought and renovated by some merchants from the guild. In 1784, the rebuilt Yuyuan Garden was opened to the public as the Western Garden of Town God's Temple of Shanghai.

In the early 20th century, Yuyuan Garden was divided into southern and northern parts by Yuyuan Road. The southern part included Huxin Pavilion, Zigzag Bridge, Exquisite Jade Rock, Moon-Embracing Pavilion, Xiangxue Hall and other rockeries and pavilions; the northern part included Cuixiu Hall, Dianchun Hall, and House of Spring Breeze, etc. Many of the buildings were converted into folk dwellings. Ninghui Pavilion, Qingfen Hall, Pleasure Boat, Green Wave Hall were turned into restaurants, bakeries or teahouses. A bustling bazaar and a paradise for the citizens took shape.

After 1949, Mr. Chen Congzhou undertook two large-scale

betterments to Yuyuan Garden. From 1956 to 1961, Sansui Hall, Wanhua Chamber, Dianchun Hall and other areas were restored, and new attractions such as Nine Lions Pavilion, Sightseeing Tower, Yuhua Hall, and Winding Water were built. From 1986 to 1988, the renovations were mainly carried out in eastern part, like Exquisite Jade Rock, Yuhua Hall, Sightseeing Tower, and Nine Lions Pavilion and so on. Huanlong Bridge was rebuilt. Jiyu Rockery, Huanyun Rockery, Universal Happiness Screen Wall and Jiyu Corridor were constructed. And the ancient stage in the former Palace of Thean Hou Temple in the north of Shanghai was relocated into the garden.

Although the present day Yuyuan Garden is only half of that of the Ming Yuyuan Garden in terms of scale, Grand Rockery built in the Ming Dynasty has survived. Sansui Hall and Dianchun Hall and other ancient buildings are grand. Huxin Pavilion in the middle of Zigzag Bridge is unique. The overall garden still showcases the charm of classical Jiangnan gardens. Huxin Pavilion and its surrounding ponds used to be a famous scene when Yuyuan Garden was first built, known as "Daoyi Pavilion" or "Fushui Pavilion". Nowadays, the plane view of Huxin Pavilion looks like "T", with two polygonal wings. The roof is composed of six tapered and short hip-and-gable roof structures of different sizes with 28 corners, in picturesque disorder and very lifelike.The Zigzag Bridge used to have wooden railings on both sides. They were replaced by cement railings during the reconstruction in 1924.

# 中西交汇

Chinese – Western

# 外滩建筑群

Building Clusters on the Bund

黄浦区、虹口区，上海大厦至延安东路口
Huangpu District, Hongkou District,
from Broadway Mansions to East Yan'an Road Intersection

一千多米长的外滩拥有中国最大的近代建筑群，涵盖了外廊式、新古典主义、哥特复兴、装饰艺术等各种风格，几乎是一部生动的缩略版近现代建筑史，更是上海的“城市会客厅”。

外滩地区现存的建筑风貌被称为“外滩三期”，主要成形于20世纪20年代。在这铅灰色西洋建筑群中，也不乏中国元素灵动的影子。2015年4月，外滩荣列住建部和国家文物局公布的第一批30个中国历史文化街区，是上海唯一入选的街区。

从外滩延伸出来，纵向有隔江而望的陆家嘴摩天大楼群，横向有沿江而去的45公里滨江岸线公共空间，在城市高度与宽度上遥相呼应并延续着百年的建筑传奇。今天的外滩，是上海当之无愧的文化地标。当你阅读每一栋建筑的故事，你就开始了解这里，也不断感受到这座生生不息的城市的温度。

The Bund stretches more than 1,000 meters along the bank of the Huangpu River, and houses dozens of historic buildings displaying various architectural styles, including veranda, neo-classical, Gothic Revival, Art Deco and many other styles. Being the richest collection of modern architecture in China, it is a vivid miniature of modern architectural history, and is reputed as the "City Parlour" of Shanghai.

The existing architecture landscape along the Bund is called the Bund Phase III, which mainly took shape in the 1920s. Traces of vivid Chinese elements are also embodied among the leaden-grey stretches of Western-style architectural complex. In April 2015, the Bund was listed by the Ministry of Housing and Urban-Rural Development and State Administration of Cultural Heritage in the first batch of 30 Chinese Historical and Cultural Blocks . It is the only block selected in Shanghai.

The Bund bears witness to the emergence of the 45-km waterfront public space along the Huangpu River and the modern skyscrapers of Lujiazui on the opposite bank, which are horizontal and vertical continuations of the century-old architectural legend. Nowadays, the Bund has become a well-deserved cultural landmark of Shanghai. Reading the stories of these old buildings, you come to understand the history of the Bund and the vast heritage of this dynamic city.

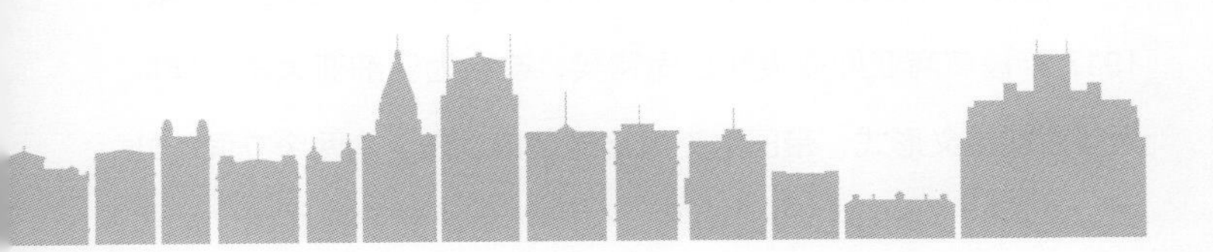

# 亚细亚大楼
Asiatic Petroleum Co. Building

亚细亚大楼位于外滩 1 号 ,1916 年竣工。初名麦边大楼，1917 年被英商亚细亚火油公司购买，改名为亚细亚大楼。建筑为新古典主义形式，带巴洛克风格的装饰。东、南两个立面的处理完全一样，均作三段式构图，东南转角墙面处理成内凹的弧形。外墙为花岗石饰面。底部两层为基座，入口处饰爱奥尼亚式双柱，上为巴洛克式的弧形断檐门楣；三至五层为中段，中部有三孔券窗；六至七层为上段，中部设贯通两层的爱奥尼亚式双柱廊，层间有铁栏弧形阳台，两侧窗楣饰三角形断山花，屋顶建有角楼和露天平台。今为久事国际艺术中心等单位使用。

Built in 1916, the Asiatic Petroleum Co. Building is located at No.1 on the Bund. Originally named the McBain Building, it was bought by the Asiatic Petroleum Co. in 1917 and renamed Asiatic Petroleum Building. Its neo-classical style is complemented by Baroque decorations. Its east and south facades, connected by a concave wall, look exactly the same: both divided into three parts, and decorated by granite. The bottom two floors constitute the pedestal. At each of the two entrances stand two Ionic-style columns, upon which lay a Baroque arched lintel. From the third to the fifth floor is the middle part, the connecting piece of its flanks has three arched windows. The upper part includes the sixth and seventh floors. In the middle are two sets of Ionic dual-column linking two floors. On each floor lies a curved balcony with iron bars. Each lintel of window at two flanks on the sixth floor is decorated with a triangular broken pediment. On the roof of the former Asiatic Petroleum Co. Building erect four turrets and a big terrace. Today the building is used as Jiushi International Arts Center and other enterprises.

# 上海总会大楼

Shanghai Club

上海总会大楼又称英国总会，位于外滩 2 号，建于 1864 年，现存建筑已是总会的第二代建筑，建于 1911 年，钢筋混凝土结构，英国新古典主义形式，有巴洛克风格的装饰。屋顶的双塔楼构图除了受巴洛克艺术的影响之外，其穹顶造型又受到印度建筑的影响。大厅内有八对贯通两层的托斯卡纳式双柱，回廊在柱间向厅内作弧形出挑，大厅顶部为拱形玻璃天棚。底层南侧酒吧有当时世界少见的 30 多米长的柚木吧台。1949 年后，这里改为上海百货批发站使用。1956 年后改为国际海员俱乐部。1971 年改为东风饭店。2009 年起改为华尔道夫饭店。

Located at No. 2 on the Bund, the Shanghai Club, also known as the British Club, was established in 1864. The tower we see now, was built in 1911. It is the second generation of the Club. With a reinforced concrete structure, it is in a British neo-classical style with Baroque decoration. The twintower structure on the roof is influenced by Baroque art, while its dome is affected by Indian architecture. Eight pairs of Tuscan dual-column linking two floors stand in the hall. A curved corridor lies towards the hall between columns, and the ceiling of the hall is an arch-shaped piece of glass. The bar in the south of the ground floor has a 110-foot long teak bar counter, a rare example of this kind in the world. After 1949, it was used as Shanghai Wholesale Station, and after 1956 started to be used as the International Seamen's Club. In 1971, it was renovated to be the state-owned Dongfeng Hotel. In 2009, it was renovated again for the Waldorf Astoria Hotel.

# 汇丰银行大楼

HSBC Building

汇丰银行大楼位于外滩 12 号，1923 年落成，底层中央设有三孔半圆拱券门廊，左右有一对铜狮，原件已移至上海历史博物馆内展出，现大楼内的为复制品。二至四层中部有贯通三层的科林斯巨柱式柱廊，顶部为三层高仿罗马圣彼得大教堂的穹顶。

主入口内是一座八角形门厅，门厅的拱肩、鼓座和穹顶上均饰有马赛克镶嵌画，极具艺术价值。汇丰大楼因其显著的位置、宏大的体量、精美的设计，而赢得美名。1955 年起为上海市人民政府使用。1996 年后为上海浦东发展银行使用。

Located at No. 12 on the Bund, the HSBC Building was completed in 1923. In the centre of its ground floor lies a porch with three semicircular arches, with a pair of bronze lions on its left and its right side respectively. The original bronze lions have been moved to and are displayed in the Shanghai History Museum. The lions now on site are replicas. From the second to the fourth floor, there is a corridor with Corinthian Order columns which penetrate the three floors in the middle. At the top is a three-storey dome imitating the Basilica of St Peter's in Rome. An octagonal hall lies at the main entrance with mosaic paintings showing classical characters and stories on spandrels, drums and domes. These paintings are of highly artistic value. The HSBC Building is reputed for its prominent location, majestic size and exquisite design. It once was the seat of the Shanghai Municipal Government building after 1955 and now is where Shanghai Pudong Development Bank located since 1996.

# 江海关大楼
Custom House

江海关大楼位于外滩 13 号，1925 年兴建。建筑分东、西两部分，东部面临外滩，西部延伸至四川中路。东部立面为三段式构图，一至二层设为基座，外墙由粗石砌筑，入口处为典型的希腊多立克柱式门廊，以及古典风格的铜质大门；中段为标准层，有细琢的花岗石饰面，强调竖向线条，七层上方挑出带齿饰的高大檐口；顶部设角亭和层层收分的四面钟楼，为典型的装饰艺术风格。室内底层大厅中央有铸铜灯座，漫射光照亮顶部的八角形藻井及彩色马赛克镶拼的帆影海事图案。顶部的大钟为中国最大、建造最早的海关大钟，其规模当时为亚洲之最。今为上海海关大厦。

Located at No.13 on the Bund, the Custom House was built in 1925. It has two sections: the east wing faces the Bund while the west wing extends to Middle Sichuan Road. The east facade features three parts with the first two floors as its base. Exterior walls are made of rough stones. At the entrance, the porch is supported by typical Doric columns, and the bronze door is classical. In the middle are standard floors with fine-carved granite finishes which emphasize vertical lines. Above the seventh floor, tall cornices with dentils stick out. A tetrahedral clock tower is in typical Art Deco style with a corner pavilion at its top and varied pillars. In the centre of the hall on the ground floor stands a cast bronze lamp holder. Beaming light illuminates the octagonal caisson ceiling and colourful mosaic maritime patterns. The clock at the top of the Custom House is the largest and earliest custom clock in China. It was also the largest of its kind in Asia at that time. Today the building is used as Shanghai Custom House.

# 汇中饭店大楼

Palace Hotel

汇中饭店大楼位于外滩 19 号，1906 年兴建，高六层，砖混结构，平面矩形。大楼在南京东路转角处，为英国文艺复兴时期的建筑风格。底层以花岗石砌筑，上部为红白间砌砖墙，南立面原有新艺术运动风格的铸铁外廊，今已不存。屋顶原设有上海最早的空中花园，东侧南北各有一座巴洛克式塔亭，造型迥异，原建筑 1912 年毁于火灾，现为 1998 年重建。楼内有上海第一台载人电梯，底层大厅为柚木装修，护墙板、柱身、楼梯及栏杆等处均有精美木雕，东部为可容纳 300 人的餐厅。今为和平饭店南楼。

Located at No. 19 on the Bund, the Palace Hotel is a six-storey building with a brick-concrete structure and rectangle shape in plane view. It was built in 1906 and designed in a Queen Anne Renaissance style. Its ground floor is entirely constructed in granite and exterior walls of upper stories feature red and white bricks. The south façade originally had a cast iron side corridor of the Art Nouveau style, which no longer exists. The first hanging garden in Shanghai once lied on the roof. On the east wing of the building, one Baroque tower pavilion stood in north and another in south. The two featured different styles and were destroyed by fire in 1912. The current tower pavilions were rebuilt in 1998. The building is equipped with the first manned elevator in Shanghai. Its hall on the ground floor is decorated with teak. There are exquisite wood carvings on dado rails, pillars, staircases and handrails. The east section is a 300-seat restaurant. Today it is south wing of the Fairmont Peace Hotel.

# 沙逊大厦
Sassoon House

沙逊大厦位于外滩 20 号，1929 年落成。建筑依道路布局，平面呈 A 字形，钢框架结构，装饰艺术风格。立面以竖向线条构图为主，檐部和基座线脚等有几何图形装饰。塔楼冠以十多米高的方锥体屋顶，外覆墨绿色瓦楞紫铜板，塔顶再设一座方尖碑造型的采光塔，塔尖今已不存。室内的大理石地面、墙面拼花图案、栏杆、天花和灯具等设计也为装饰艺术风格。底层有两条穿过式拱廊，中间交汇点有八角形大厅。内有著名的九国风格客房。沙逊大厦是标志着上海近代建筑终结复古样式走向现代的里程碑，是当时外滩最豪华的一幢高层建筑，曾有“远东第一楼”之称。沙逊大厦于 1956 年后改为和平饭店。

Located at No. 20 on the Bund, the Sassoon House was completed in 1929. The layout of the building was designed following the roads with an "A" shape in plane view. It was built in Art Deco style with a steel framework. The facades are mainly composed of vertical-line designs, and the entablatures and base mouldings are decorated with geometric figures. The tower roof is a square-based pyramid which is more than 10 metres high and covered with dark green corrugated copper plates. At the top of the tower also stands an obelisk-shaped lantern and now the spire has gone. Indoor marble grounds, walls in mosaics patterns, and design of railings, ceilings and lamps are also in Art Deco style. On the ground floor, there are two arcades. There is an octagonal hall at the meeting-point in the middle. There are famous suites with nine different country styles in the former Sassoon House. The building marks the end of retro form and the start of modern style in the history of modern architecture in Shanghai. It was the most luxurious high-rise on the Bund, and was known as "the first building in the Far East". The Sassoon House was renamed the Peace Hotel after 1956.

# 中国银行大楼

The Bank of China Building

中国银行大楼位于外滩 23 号，建于 1935—1944 年。建筑装饰艺术与中国传统风格相结合。这座西式摩天楼有着琳琅满目的中式元素，让人惊叹中西文化的水乳交融。外墙为花岗石饰面，立面强调垂直线条和几何图案装饰，顶部两侧呈台阶状，塔楼部分冠以平缓的铜绿色琉璃瓦四方攒尖顶，檐部施以石斗拱装饰，每层两侧配以镂空花格窗，大门上方原饰有孔子周游列国石雕，栏杆花纹及窗格也采用中国传统形式。门外有九级石阶，两扇紫铜图案雕饰大门。营业大厅内有大理石饰面的列柱和柜台，顶部为拱形玻璃天棚。大楼是外滩建筑群中唯一具有中国特色的早期现代高层建筑。今为中国银行上海分行所在地。

The Bank of China Building is located at No. 23 on the Bund. Built from 1935 to 1944, it combines Art Deco style and traditional Chinese architectural features. Looking at this Western-style skyscraper with Chinese elements, people cannot help but marvel at the chemistry between Chinese and Western culture. Exterior walls are decorated with granite, while the facades highlight vertical-line designs and geometrical decorations. Both sides of the top are step-shaped. At the top of tower lies a gentle pyramid roof made of copper green glazed tiles. The entablatures are decorated with stone brackets, both sides of which are equipped with hollow-out lattice windows. Over the gate there was once a stone carvings showing Confucius travelling around the countries. The railing patterns and window lattices also follow traditional Chinese style. There are nine steps outside the building. Its two gates are decorated with copper patterns. Colonnades and counters decorated by marble stand in the business hall and the ceiling is covered by arched glass. It is the only early modern high-rise with Chinese characteristics among the building clusters on the Bund. Today, it is used as Bank of China Shanghai Branch.

# 怡和洋行大楼
Jardine Matheson Building

怡和洋行大楼位于外滩 27 号。钢筋混凝土结构，欧洲古典复兴风格。立面三段式处理，一、二层外墙采用粗凿花岗石垒砌，立面作连续券廊构图，券内门窗上下通连；三至五层外墙用磨琢较细的花岗石垒砌，立面作柱廊构图，东立面中部五间，北立面中部四间，均贯以变体科林斯式柱，顶部旗杆周围原有石雕海豚像，后因加层被拆除；现上部加建的两层采用仿石饰面层，屋顶设有两米多高的石壁，中竖旗杆。怡和洋行大楼是上海采用石料外墙的早期实例。今为罗斯福公馆等单位。

The former Jardine Matheson Building is located at No. 27 on the Bund. It is rectangle shaped in plane view. The building has a reinforced concrete structure and is in the European classical renaissance style. The building facades are divided into three parts from top to bottom. Exterior walls of the first and second floors feature rough-hewn granite and the facade has serial arcades with connecting doors and windows inside. From the third to the fifth floors, the exterior walls are built with fine granite and the facade has colonnades with five in the middle of east facade and four in the north. They all have Corinthian columns. There were originally stone dolphin statues around the top flagpole, but they were removed due to the addition of stories. Now the upper two additional floors adopt stone-like facing layer. The roof has two-metre high stone walls and the flagpole erects in the middle. The Jardine Matheson Building is an early example of using stone exterior walls in Shanghai. Today, it is used as the House of Roosevelt and other enterprises.

# 东方汇理银行大楼

Banque de L'Indo-Chine Building

东方汇理银行大楼位于外滩 29 号，建于 1912—1914 年。法国古典主义风格，带有巴洛克装饰。立面作三段式划分，比例严谨，底层门窗处理成高大的拱券，入口处有一对托斯卡纳式柱，门楣饰巴洛克式卷涡状断山花。二、三层中部有贯通两层的爱奥尼亚式柱，二层窗户带有柱式和拱券装饰，其正中的窗户采用帕拉第奥式组合，顶部带有精致的垂花雕刻。室内以大理石装饰，营业大厅为双排爱奥尼亚式柱廊，玻璃拱顶，是当时银行建筑中较为通用的格局。东方汇理银行大楼是外滩建筑群中一座十分精致的建筑。今为光大银行上海外滩支行使用。

Located at No. 29 on the Bund, the Banque de L'Indo-Chine Building was built from 1912 to 1914. Featuring French classical style with Baroque decoration, its facades are divided into three parts and meticulous in layout. The doors and windows on the ground floor are in tall-arch-style with a pair of Tuscan columns at the entrance and Baroque circinate broken pediments on door frames. In the middle of the second and the third floor are Ionic columns linking the two floors. Windows on the second floor have column and arch decorations, among which the middle one is in Palladian style and with exquisite festoon carvings. The interior is decorated with marble. Double rows of Ionic columns stand in the business hall with a glass vault at the top. These were common patterns for bank at that time. It is a very exquisite building among the buildings on the Bund. Now it is usesd as China's Everbright Bank Shanghai Branch.

# 百老汇大厦

Broadway Mansions

百老汇大厦位于北苏州路 20 号，1934 年建成。建筑坐北朝南，平面呈 X 形，双层铝钢框架结构，装饰艺术与美国现代高层建筑风格相结合。立面中间高、两侧低，呈跌落式构图，所有的顶部檐口均饰以统一的几何形连续装饰图案。外墙贴棕色面砖，窗裙部分拼成图案。大厦建成时是专供在上海的英国商务人员短期居住的公寓。1951 年后改名为上海大厦。

The Broadway Mansions is located at 20 North Suzhou Road and was established in 1934. Facing south, the building is an X-shaped double aluminium and steel frame structure, combining art-deco style with modern American high-rise style. The facade is tall in the middle and has setbacks layer upon layer on both flanks. All cornices at the top are decorated with continuous geometric patterns. Exterior walls are covered with brown tiles which form patterns in the windows. The mansions used to be residence for British businessmen in Shanghai for a short period of time. After 1951, it was renamed Broadway Mansions Hotel Shanghai.

上海市历史博物馆

Shanghai History Museum

上海市历史博物馆建筑的前身是跑马总会大楼，总会成立于1850年，其跑马场前后经三次变迁。1860年，又在今人民广场、人民公园一带开辟第三跑马场，习称跑马厅。

第三跑马场有草地跑道和硬跑道两个跑圈。外圈的草地跑道全长约2000米，用于赛马；里圈的泥质跑道及中间的大片场地，场地上曾先后建有游泳池、板球场、高尔夫球场和足球场等设施。

跑马厅每年举办赛马、开彩，遂在跑马厅内建一幢大楼，成立高级俱乐部。1932年，总会拿出200万两银子，在跑马厅西北角造楼。大楼占地面积890平方米，建筑面积2100平方米，高四层。外观是古典主义构图，兼具折衷主义，外墙用红褐色面砖与石块，二、三层西立面贯以托斯卡纳式柱廊。西北端高耸53.3米的钟楼，最上方是四面三角形坡形顶，顶与大钟之间是瞭望台。一、二层与大看台相连，看赛马则在三楼长廊中。底层设售票处、领奖处。一、二层间夹层为滚球场。二楼有咖啡室、游戏室、弹子房、阅览室等。三层有会员包房、餐厅，四层是职工宿舍。1925年、1930年及1935年，大楼南面陆续造了具有英国早期近代建筑风格的红砖墙二层房屋，直至跑马厅路（今武胜路）口。

1949年上海解放，改建为人民公园、人民广场。1952年，先后被用作上海博物馆和上海图书馆。两单位先后迁出，改由上海美术馆使用。1989年被公布为上海市文物保护单位。现为上海市历史博物馆、上海革命历史博物馆，于2018年3月正式对外开放。

黄浦区南京西路325号，原跑马总会
325 West Nanjing Road,
Huangpu District, Former Shanghai Race Club

Shanghai History Museum occupies Shanghai Race Club. The Shanghai Race Club was founded in 1850. It changed location three times. In 1860, the property was sold again. The owners bought a piece of land triple its former size and constructed the third racecourse, commonly known as the Shanghai Racecourse.

The third racecourse had two tracks: one turf track and one hard track. The outer turf track, around 2,000 metres long, was used for horse racing; and the inner clay track and the large field inside once hosted a series of facilities, including swimming pool, cricket pitch, golf course and football field.

The Shanghai Racecourse held horse races and lotteries every year, so some board members suggested erecting a building inside the Shanghai Racecourse for an exclusive club. In 1932, the club spent two million taels of silver in constructing a building at the north-western corner of the Shanghai Racecourse. It covered an area of 890 square meters with a floor area of 2,100 square meters, standing four floors tall. It was in the classic style with a touch of eclecticism. The façade was covered with a combination of brown face bricks and stone blocks, and Tuscan columns spanned the western facades of its second and third floors. At its north-western corner was a 53.3-meter-high bell tower topped by a pyramid-shaped slope roof. Between the roof and the bell was the observatory. The first floor and the second floor were connected to the grandstand, and club members would watch the races from the corridor on the third floor. The ticket office and the prize-collecting office were on the ground floor. The mezzanine between the first and second floors was used as a bowling alley; the second floor as a members' club, complete with a cafe, a game room, a billiard room, a reading room, etc.; the third floor as VIP rooms and restaurants; and the fourth floor as staff quarters. In 1925, 1930 and 1935, respectively, the club built two-storey houses of early modern British style with red bricks, which extended all the way to the intersection of Race Course Road (now Wusheng Road), on the south of the original building.

After Shanghai was liberated in 1949, it was rebuilt into People's Park and People's Square. In 1952, the building of the Shanghai Race Club was used as the Shanghai Museum and Shanghai Library. The two organizations moved out in succession, and the building became the Shanghai Art Museum.In 1989 it was designated as a cultural site protected by the city of Shanghai. Now it has been transformed into the Shanghai History Museum and the Shanghai Revolution Museum, which officially opened to the public from March 2018.

# 上海科学会堂

Shanghai Science Hall

黄浦区南昌路47号
47 Nanchang Road, Huangpu District

上海科学会堂与复兴公园比邻，都是上海的地标性建筑，知名度颇高。2019 年，上海科学会堂的老洋房被公布为全国重点文物保护单位，受到了更多人的关注。

1904 年，法国商会及社会名流组织了一个名为 Cercle Sportif Français 的社会团体，在中国人所称“顾家宅花园”或“法国花园”，即今天复兴公园的北侧建造会所，即法国总会。1917 年在公董局法国建筑师万茨和博尔舍伦主持下对法国总会建筑进行扩建。扩建工程原则上保留原来的建筑，向西加建新建筑，使建筑东西长度达到 130 余米。扩建后的主楼以中央为中轴线，两翼对称，中轴线加建钟楼和前置的露台，两边的屋面盔式折坡，即所谓“孟莎式屋顶”，比原来的立面更气派、更豪华。法国总会的出入口设在环龙路（今南昌路）上，进门就是宽敞的扶梯，通往左右两侧，扶梯有铸铁栏杆，是用法文字母 C.S.F. 组成的图案，是 Cercle Sportif Français 的缩写；出入口面对朝南的大彩绘玻璃窗，阳光透过玻璃射入室内，光线变得柔和温暖，营造出一派愉悦、祥和的气氛；在北侧的彩绘玻璃上还能找到法文“L’Orphelinat de Tou-sè-wè 1918”（土山湾孤儿院），这大概是目前所知上海唯一有标记和制造年份的土山湾孤儿院生产的彩绘玻璃，文物价值极高。中轴线两侧的底层和二层设计有宽敞的走廊，分别设有开仑、斯诺克、保龄球房、酒吧间、舞池、击剑房、更衣室、大餐厅等，法国总会与英商上海总会、德国总会、美国总会合称“四大总会”。

后来法国人在迈尔西爱路（今茂名南路）又建造了一个法国总会，于是，公董局就把环龙路的法国总会部分作为公董局学校和法国人学校。

1949 年后，原来南昌路的法国总会曾成为上海市文化局办公用房，1957 年改为科技工作者活动场所。1958 年，上海市科学技术协会成立，这里成为其会址。

The Shanghai Science Hall stands next to Fuxing Park, both popular landmarks of Shanghai. In 2019, the old Western-style building of the Shanghai Science Hall was declared a Major Historical and Cultural Site Protected at the National Level, gaining more attention.

In 1904, the French Chamber of Commerce and public celebrities established a social group named Cercle Sportif Français and built the "French Club" in the north of what the Chinese called the "Gujiazhai Garden" or "French Garden" now known as Fuxing Park. In 1917, French architects Wantz and Boisseron of the Municipal Administrative Council of the French Concession undertook the expansion project of the French Club. In principle, the expansion project retained the original building, and new buildings were built to the west, extending the total length of the building to more than 130 metres from east to west. The expanded main building takes the central as the central axis, with two wings in symmetry. In the central axis, a bell tower and a front terrace were added. The roofs on both sides are helmet-shaped folding slopes, the so-called "mansard roof," which is more magnificent and luxurious than the original façade. The entrance to the French Club is on Route Vallon (today's Nanchang Road). Two staircases on both sides lead to the second floor, with cast iron railings carved with the patterns of French abbreviation C.S.F. (Cercle Sportif Français); across the entrance are large stained-glass windows facing south. The sunlight shines into the room through the windows, and the soft and warm light creates a pleasant and peaceful atmosphere; the French words "L'Orphelinat de Tou-sè-wè 1918" can still be found on the stained glass on the north side. This is probably the only existing stained glass manufactured by the Tushanwan Orphanage marked with its logo and production date in Shanghai, and thus has a very high cultural value. The first floor and the second floor on both sides of the central axis are designed with spacious corridors, facilitated with a carom room, a snooker room, a bowling room, a bar room, a dance floor, a fencing room, a dressing room, a restaurant, etc. The French Club, together with the English Club, the German Club and the American Club in Shanghai are collectively called the "four great clubs".

Later, the French built a new French Club on Route Cardinal Mercier (today's South Maoming Road). Therefore, the Municipal Administrative Council of the French Concession took the French Club on Route Vallon as its school and a school for French people.

After 1949, the French Club on Nanchang Road was used as the office of the Shanghai Municipal Administration of Culture. In 1957, it was changed into an activity centre for scientific and technological workers. In 1958, the Shanghai Association for Science and Technology was established and this place became its seat.

# 工部局旧址

Former Site of Shanghai Municipal Council

黄浦区汉口路193号
193 Hankou Road, Huangpu District

上海市中心区汉口路、江西中路、福州路、河南中路相围的那块土地大致上呈 140 米见方的正方形，面积约两公顷。这里曾经是上海市政府机关所在地，而在此之前，它是上海公共租界工部局机关楼，故称“工部局大楼”。

随着租界经济发展，人口增多，面积扩大，工部局的权力越来越大，机构也日益齐全和庞大，原来的办公场所就不敷使用了。

1913 年，新屋的方案和建筑设计方案和图纸送到英国，请英国皇家建筑师学会审查，几经修改。1914 年正准备破土动工时，第一次世界大战爆发，工程被迫终止，一直到 1918 年战后重新启动。工程于 1922 年 11 月 16 日才正式交付使用。

工部局大楼的建筑平面呈“回”字形，以朝东北的汉口路立面为主要立面，江西路和福州路为次立面，新古典主义风格，四层钢筋混凝土结构，门窗大多使用石材，外墙则以加工后的石材作为贴面，从视觉上感觉这是一幢用石头垒起来的城堡。

主入口设计在汉口路及其两端，其中汉口路与江西路转角的东北入口为大楼的主入口，设计为花岗石多立克双柱支撑的扇形门廊，小汽车可以直接驶入门廊，入门后为宽畅的厅，左右均设计有大理石的楼梯，墙面也使用大理石装饰，十分气派；门廊上改为平台，该入口的上方二层就是工部局总董的办公室，以及董事会议室。每层有很宽很长的走廊，两边均为办公室。工部局下设的各处室全部集中在该大楼内。

1941 年 12 月 7 日太平洋战争爆发，次日，工部局实际上已被日军掌控，1943 年年底，汪伪上海特别市政府迁入大楼。1945 年抗战胜利后，这里成了国民政府的上海市政府。

1949 年至 1956 年，这里曾经是上海市人民政府驻地。1989 年被公布为上海市文物保护单位。

The plot surrounded by Hankou Road, Middle Jiangxi Road, Fuzhou Road and Middle Henan Road in the central district of Shanghai is roughly a square, with an area of about two hectares. It used to be the seat of the Shanghai Municipal Government, and before that, it was the office building of the Municipal Council of the Shanghai International Settlement.

With the economic development of the International Settlement, the population grew, the area expanded, and the Municipal Council also had more and more power, and the growing bodies were increasingly complete and expanded, thus the original office space was no longer big enough for use.

In 1913, the plan and architectural design of the new building were sent to the UK to be reviewed by the Royal Institute of British Architects, and was revised for several times after that. In 1914, just when the construction was about to start, World War I broke out, and it was forced to terminate. It was not until 1918 after the war did the construction restart. On November 16, 1922, the building was officially put into service.

The building plan of the Shanghai Municipal Council building is in the shape of a small square within a bigger one. The main facade is on Hankou Road facing northeast, and the side facade are facing Jiangxi Road and Fuzhou Road. The four-storeyed building is reinforced concrete structure of neoclassical style, with most of the doors and windows made of stones and the exterior walls veneered with processed stones. The building seems just like a castle of stone form outside.

The main entrances are designed on Hankou Road and its two ends. The northeast entrance at the corner of Hankou Road and Jiangxi Road is the major entrance to the building, with the fan-shaped porch supported by twin Doric columns of granite. Cars can drive directly into the porch. Inside the door, there is a spacious hall, where marble stairs are designed on the left and right, and the walls are extravagantly decorated with marble. The porch top is a terrace. The office of the managing director of the Municipal Council and the board room are on the second floor above the entrance. Each floor has a very wide and long corridor with offices on either side. The offices under the Municipal Council are all in the building.

On December 7, 1941, the Pacific War broke out. The next day, Shanghai Municipal Council was actually under the control of the Japanese army. At the end of 1943, the puppet Shanghai Special Municipal Government moved into the building. After the victory of the War of Resistance against Japanese Aggression in 1945, it became the Shanghai Municipal Government of the Nationalist Government.

From 1949 to 1956, this place was the station of the Shanghai Municipal People's Government. In 1989 it was designated as a cultural site protected by the city of Shanghai.

# 上海邮政总局

Former Shanghai General Post Office Building

虹口区北苏州路276号
276 North Suzhou Road, Hongkou District

上海是中国最早兴办近代邮政的城市。

1872 年，海关总税务司建议，仿英国的邮政制度，以海关邮驿为基础成立“寄信局”，在有条件的地方开辟全国邮政，成为中国邮政之始。

1896 年 2 月 19 日，清廷宣布设立总邮政司，受制于总理衙门，并任命总税务司赫德兼任总邮政司，第二天，设在上海海关内的寄信局即改组为大清邮政局，同时将全国各地划分为 35 个邮政支局或邮界。中华民国建立后，大清邮政局又改组为中华邮务总局，机构庞大了，原所在建筑不敷使用，遂于 1922 年计划择址重建，局址选在与北火车站和黄浦江码头水陆交通都便利的四川路桥北堍。该大楼于 1922 年 12 月动工兴建，1924 年 12 月建成交付使用。

大楼由英商思九生洋行设计。建造大楼的土地分两次买进，总面积 9.727 亩，建筑面积 25294 平方米，建筑高度 51.16 米 ( 不包括旗杆 )。以北四川路与北苏州河路相交处为第一立面，两翼展开，大致对称。主立面设计为带巴洛克风格亭式顶的钟楼，入大门后，两侧有环状的大理石扶梯，将客人直接引入二楼，这里是一个面积约 1200 平方米的营业厅，以大理石铺地，两边均为大理石柜台，柜台上装有精致的铜质栏杆，大厅十分气派而华丽。展开的两翼均设计为通层的科林斯柱支撑的玻璃墙面。

值得一提的是主立面的钟楼和雕像。这座钟楼分作三层，正面和背面均为平面，两边呈圆形，边角设计有明显凹凸的墙隅石，檐部也有浮雕，明显增加了建筑物的阴暗面和立体效果，从而更突出钟面；钟楼高 13 米，东西两旁各有一火炬台；钟楼上面是钟塔，高 17 米，呈四边形，并附有雕刻精细的弧形栏杆，弯曲的口和塔顶。塔身四面有门框，门两边有成对的爱奥尼亚式双柱。塔顶有旗杆，杆高 8.2 米。塔基的两旁各有一组三人组成的雕塑群像，一组的三人手中各持火车头、轮船铁锚和通讯电缆的模型。另一组三人中，居中者是赫耳墨斯，是古希腊商业之神，立像作站立握钱袋状；两边分别为爱神厄洛斯和阿佛洛狄忒，这组群雕寓意，邮政是信使，传递信息，沟通人们之间的情感。

1996 年，上海邮政总局被公布为全国重点文物保护单位。今为上海邮政博物馆等单位使用。

Shanghai was the first city in China that pioneered modern postal service.

In 1872, the Inspector General of the Maritime Customs Bureau suggested setting up a "head post office" by referring to the British postal system and on the basis of the customs post offices, and to opening national postal service in places where conditions permitted.

On February 19, 1896, the Qing government announced to set up the Head Post Office, which was subject to the Office of Foreign Affairs, and Sir Robert Hart, the then Inspector General of the Maritime Customs Bureau, was appointed to be the head of it. The next day, the post office in Shanghai Customs was reorganized as the Great Qing Imperial Post Office. 35 postal branches or postal offices were set up all over the country to establish China's postal network. After the founding of the Republic of China, the Great Qing Imperial Post Office was reorganized into the Chinese Post Office. The previous premises of the postal administration was no longer adequate, and in 1922, a land for the construction of a headquarter building was acquired at the north end of the Sichuan Road Bridge, where the water and land transportation were convenient with the North Railway Station and Huangpu River Wharf in the neighbourhood. It broke ground in December 1922, and was completed and put into use in December 1924.

The building was designed by the British company Stewardson & Spence. It is 51.6 meters high, covering a total area of 9.727 mu, and has a floor area of 25,294 square meters. The main facade was at the intersection of North Sichuan Road and North Suzhou River Road with two wings unfolded and roughly symmetrical. The main facade is topped by a Baroque style clock tower. Facing the main entrance, two marble treaded circular stairways on both sides lead to the 1200 square meters main trading hall on the second floor. Marble paves the floor, and on both sides stand marble counters, which feature exquisite copper railings. The hall is magnificent and gorgeous. Both wings use glass walls and three-storey high Corinthian order columns.

The things worth particular mentioning are the bell tower and sculptures on the main facade. The bell tower is divided into three floors. The front and back are flat, and both sides are circular. On the corners are remarkably scrabrous quoins, while the reliefs on the eaves significantly increase the building's shadow and three-dimensional effect, highlighting the clock face; the bell tower is 13 meters high, with a torch platform on either of the east and west sides; on top of the bell tower is a 17-meter-high square-shaped clock tower with finely carved arc railings, a curved mouth and a tower top. There are door frames on four sides of the tower, and there is a pair of Ionic order columns on both sides of a door. There is an 8.2-meter-high flagpole on top of the tower. On each side of the tower base is a three statuary group. On one side, the sculptures respectively hold models of locomotive, ship anchor and communication cable. The other group features Hermes, the God of Commerce of ancient Greek, standing with a purse in hand; Hermes is flanked by Eros and Aphrodite, the god and goddess of love. This group of sculptures is an indication that postal service is like a messenger, transmitting information, communicating people's feelings.

In 1996, The building was announced as a Major Historical and Cultural Site Protected at the National Level. It is now used by Shanghai Postal Museum and other enterprises.

# 提篮桥监狱早期建筑
Early Buildings of Tilanqiao Prison

虹口区长阳路147号
147 Changyang Road, Hongkou District

提篮桥监狱，全称为“上海公共租界工部局警务处监狱”，俗称西牢或外国牢监。最早由工部局创建于 1901 年，1903 年投入使用。最初的主要建筑包括两幢四层的监楼，囚室 480 间，以及炊场、办公楼、医务所等，占地面积约十亩。1916 年起，先后向北、向东扩建，到 1935 年基本定型。现存建筑物均为 1917 年至 1935 年间建造。占地面积 60.4 亩，有十幢四到六层的监楼，将近 4000 间囚室，以及工场、医院、炊场、办公楼等建筑。监狱规模宏大、建筑优良，且收押犯人数量多，曾号称“远东第一监狱”。

提篮桥监狱在结构上，采用“三墙一栅”，即三面系钢骨水泥墙，一面为铁栅，囚室全部建在有通风和敞开状的建筑内；在监区设计上，包括男犯监区、女犯监区、外籍犯监区和儿童感化院，分押分管的现代监狱管理制度特征明显。

提篮桥监狱 1951 年更名为“上海市监狱”，1995 年更名为“上海市提篮桥监狱”。监狱中的一幢十字形楼房改建成上海监狱陈列馆，于 1999 年《监狱法》颁布实施五周年之日开馆。

1994 年，提篮桥监狱被列为上海市优秀近代建筑保护单位，2013 年被公布为全国重点文物保护单位。

The Tilanqiao Prison, with the full name of "Prison of Shanghai Municipal Police Force", is commonly known as Western Prison or Foreign Prison. It was first established by the Shanghai Municipal Council in 1901 and was put into service in 1903. The original main buildings include two four-storey prison buildings, 480 cells, as well as kitchens, office buildings, medical centre, etc., covering an area of about 10 mu. Since 1916, it expanded northward and eastward successively, and basically took shape in 1935. All existing buildings were built between 1917 and 1935. Covering an area of 60.4 mu, the building complex includes ten four to six storey prison buildings, nearly 4,000 cells, as well as workshops, hospital, kitchens, office buildings and other buildings. The prison was once known as "the first prison in the Far East" for its large scale, excellent construction and large number of prisoners.

The Tilanqiao Prison adopts the structure of "three walls and one gate", that is, three sides are ferroconcrete walls, and one side is an iron gate. All of the cells are built in ventilated and open buildings; the prison zone is designed with wards for male prisoners, for female prisoners, for foreign prisoners and children's reformatory school, with evident characteristics of separate custody and management of a modern prison management system.

The Tilanqiao Prison was renamed "Shanghai Prison" in 1951 and in 1995, it was changed to "Shanghai Tilanqiao Prison". A cross-shaped building in the prison was converted into the Shanghai Prison Exhibition Hall, which opened on the fifth anniversary of the promulgation and implementation of the Prison Law in 1999.

In 1994, the Tilanqiao Prison was listed as a Monument under the Protection of the Shanghai Municipality, and in 2013, it was announced as a Major Historical and Cultural Site Protected at the National Level.

# 上生·新所
Columbia Circle

长宁区延安西路1262号，原美国乡村总会
1262 West Yan'an Road, Changning District,
Former Columbia Country Club

2018 年 5 月，延安西路 1262 号经过两年改造后开放，中文名“上生·新所”和英文名“Columbia Circle”（哥伦比亚圈）分别镌刻着两段独特的城市历史。

在沪美侨于 1917 年春成立哥伦比亚乡村俱乐部，因会员连年数以百计地增加，故在大西路（今延安西路）以南购地 48 亩筹建新所，指定美国建筑师哈沙德设计。《密勒氏评论报》1923 年刊登的手绘图纸显示，该项目拥有大型的游泳池、健身房、舞厅、壁球室、保龄球房，还有桥牌室、麻将室、烧烤吧、餐厅等，其山墙、门头、所罗门螺旋柱和细部装饰等具有典型的南加州西班牙建筑风格。哥伦比亚乡村俱乐部被用于健身娱乐，也是美侨家庭社交、节庆聚会的热闹场所。

1953 年，上海生物制品研究所迁至此处，一墙之隔的原孙科住宅成为办公楼，乡村总会的主楼作为会议、接待室，二层作为生物安全实验室，健身房成为培养基蒸锅车间，而游泳池继续为员工服务。1965 年，上海民用建筑设计院郭博主持设计了现代主义风格的高层建筑——八层高的麻腮风生产大楼，作为自主研发和生产疫苗的基地。

2016 年，上海生物制品研究所生产功能迁移至奉贤，这一处位于新华路、愚园路和衡山路—复兴路三个历史文化风貌区之间的宝贵区域成为上海万科首个城市更新项目。优秀历史建筑哥伦比亚乡村会所、上海市文物保护单位孙科旧居、麻腮风生产大楼等不同历史时期兴建的 20 栋主体建筑和园区内大量树木得以保留，生物制品生产车间时期的工业设施也融入了新的设计。见证了近一个世纪风雨的延安西路 1262 号，转型为一处全天候向社区公众开放的城市公共休闲空间。

In May 2018, 1262 West Yan'an Road opened to public after two years of renovation. Behind the Chinese name "上生·新所" and the English name "Columbia Circle" are two unique periods of urban history.

The Americans living in Shanghai established the Columbia Country Club in the spring of 1917. As the number of members increased by hundreds in successive years, they purchased 48 mu of land to the south of Great Western Road (today's West Yan'an Road) to prepare for the construction of a new club, and appointed the American architect Elliott Hazzard to design it. According to the hand-painted drawings published in Millard's Review in 1923, the project was designed with a large swimming pool, gymnasiums, dance halls, squash rooms, bowling rooms, bridges and mahjong rooms, barbecue bars, restaurants, etc. Its gables, doorheads, Solomon spiral columns and detailed decorations are characteristic of Southern California's Spanish colonial revival style.The Columbia Country Club was a centre for fitness and entertainment, as well as a venue for social and festive gatherings for American families in Shanghai.

In 1953, Shanghai Institute of Biological Products Co., Ltd. moved to this place. The former house of Sun Ke, separated by a wall, became its office building. The main building of the club served as a meeting and reception room. The second floor served as a biosafety laboratory. The gymnasium became a medium steamer workshop, while the swimming pool continued to serve the staff. In 1965, Guo Bo of the Shanghai Civil Architecture and Design Institute presided over the design of a modern style high-rise building—the eight-storey MMR production building, as a base for independent research & development and production of vaccines.

In 2016, the production base of Shanghai Institute of Biological Products Co., Ltd. moved to suburban Fengxian District, and this prime lot, hemmed in by the three Historic and Cultural Zones of

Xinhua Road, Yuyuan Road and Hengshan Road-Fuxing Road, became the first urban renewal project of Vanke in Shanghai. The 20 buildings built in different historical periods, including the Columbia Country Club, the former house of Sun Ke and the MMR production building, as well as a large number of trees in the area have been preserved. The industrial facilities of the biological products production workshop have also been integrated into the new design. Thus, 1262 West Yan'an Road, which has witnessed the vicissitudes for nearly a century, has been transformed into an all-weather urban public leisure space open to the public.

# 上海交通大学早期建筑

Early Buildings of Shanghai Jiaotong University

徐汇区华山路1954号
1954 Huashan Road, Xuhui District

上海交通大学始于 1896 年清政府创立的南洋公学，其徐汇校区是中国唯一建于 19 世纪、一所大学横跨三个世纪使用至今的大学校园。早期建筑整体布局为英美学院式平面布局——方格式格局，此形式常见于英国中世纪和美国早期的大学，如哈佛大学、耶鲁大学、哥伦比亚大学等，与中国传统建筑注重对称的布局截然不同。这也显示出交大自创办之初就立志“比肩欧美一流大学”的办学理念。

历经 120 多年，上海交通大学早期建筑仍基本完整地保留了下来，共计 16 处建（构）筑物，从 1899 年建成的中院到 1954 年的新上院，其间还有新中院、图书馆、北四楼、盛宅、体育馆、执信西斋、工程馆、总办公厅、校门、科学馆、文治堂、新建楼、五卅纪念柱，以及史霄雯、穆汉祥烈士墓。

交大早期建筑在营造、设计等方面的重要成就和价值，集合了邬达克、庄俊、范文照、杨锡镠等近现代建筑大师的作品，能够在一座校园内集中起这些经典作品，并不多见。

这些建筑如同一个展现中国近现代建筑风格变迁的博物馆，从起初安妮女王风格的中院和老上院，到折衷风格的新中院和巴洛克风格的图书馆，再到装饰艺术风格的工程馆和学院派风格的体育馆和总办公厅，以及校门的民族主义风格，大礼堂的现代主义建筑形式，新上院的社会主义民族风格……一直都跟随着时代的潮流。其中有的建筑本体已较为少见，如建于 1910 年的新中院，具有起源于印度、东南亚等地区的外廊式建筑风格，已经成为上海地区保留的为数不多的实例。

交大早期建筑中很多都采用了当时先进的建造材料和技术手段，如 1910 年建造的新中院已经在中庭安装了玻璃屋顶；1925 年建造的体育馆使用了跨度大于 20 米的钢桁架，室内还配有用锅炉调节水温的游泳池；1930 年建成的执信西斋安装了抽水马桶，是当时国内条件最好的学生宿舍。

历经时代变迁，上海交大早期建筑传递着独特的精神文化价值，构成莘莘学子的宝贵青春记忆。2019 年，被公布为全国重点文物保护单位。

Shanghai Jiaotong University (SJTU) was first established as Nanyang Public School by the Qing government in 1896. Its Xuhui campus is China's only campus used by the same university for three centuries since it was built in the 19th century. The overall layout of its early buildings is similar to that of British and American colleges, namely, the checkered layout often seen in the Middle-Age British colleges and the early American colleges, such as Harvard, Yale and Columbia. It is completely different from the symmetrical layout of traditional Chinese buildings. It was an attempt from SJTU to demonstrate its ambition to "compete with top western universities".

More than 120 years have passed by, and most of its early buildings are still kept intact. There are sixteen of them in total, from the central court built in 1899 to the new upper court built in 1954, including the new central court, the library, the No.4 North Building, Sheng's Villa, the stadium, Zhixin's Western Villa, the engineering building, the general office building, the main gate, the science building, the Wenzhi Auditorium, the New Building, the May 30th Movement Memorial Column, the Martyr Shi Xiaowen and Mu Hanxiang's Cemetery.

These buildings are significant and valuable in terms of architecture and design first because they constitute a cluster of works by Ladislav Hudec, Zhuang Jun, Fan Wenzhao, Yang Xiliu and other modern architects. It is very rare to see a campus that hosts the best works of them all.

Then, they are like a museum that displays the stylistic evolution of modern Chinese architecture, from the central court and new upper court of Queen Anne style, to the new central court of eclectic style and the library of baroque style, and to the engineering building of art deco style, the stadium and general office building of beaux arts style, the main gate of traditional Chinese style, the auditorium of modernist style, and the new upper court of socialist ethnic style. They have been evolving with the time. The prototypes that some of them are based on are very rare now, such as the new central court built in 1910. It is a veranda building whose design originated in India and Southeast Asia. It is one of the few buildings of the style still preserved in Shanghai.

Moreover, many of SJTU's early buildings adopted cutting-edge building materials and technologies at the time. For example, the new central court built in 1910 has a glass roof over its atrium; the stadium built in 1925 has a steel truss spanning for more than 20 meters and a swimming pool whose water temperature was boiler-controlled; Zhixin's Western Villa, built in 1930, was a student dormitory building equipped with flush toilets.

With the change of times, the early buildings of SJTU continue to pass on their unique spiritual and cultural value and bear witness to the youth of countless SJTU students. In 2019, it was announced as a Major Historical and Cultural Site Protected at the National Level.

# 圣约翰大学近代建筑
Historic Buildings of St. John's University

长宁区万航渡路1575号
1575 Wanhangdu Road, Changning District

圣约翰大学，这所成立最早、在华办学时间最长的教会大学，是近代上海乃至全国最著名的大学之一，2019 年被公布为全国重点文物保护单位。如今在长宁区万航渡路华东政法大学的校园内，中西合璧的建筑依旧静谧祥和，仿佛在向人们述说着那段历史。

1878 年，美国圣公会上海主教施约瑟购得 84 亩沪西梵皇渡地区的土地，并将圣公会原有的培雅书院和度恩书院合并，于次年创办圣约翰书院。1892 年开设大学课程，全面引入现代西方学科学制和校园活动。

在卜舫济校长的推动下，1902 年的书院成为圣公会私立大学，更名为圣约翰大学。1905 年在美国华盛顿注册为综合性教会大学，设文、理、医、工、神五个学院以及附属预科学校，哈佛大学、耶鲁大学等多所名牌大学皆承认其学历，毕业生可直读美国大学的研究生院，也成为中国第一所授予学士学位的大学。

八一三事变爆发后，圣约翰大学迁往公共租界，与沪江大学、东吴大学、之江大学等校组成上海基督教联合大学，1939 年迁回原址。1950 年与美国圣公会脱离关系，改由国人自办。1952 年院系调整，各学科并入上海各高校，原址成立华东政法学院。

19 世纪的圣约翰大学，仅有礼拜堂、怀施堂和科学馆，后不断扩建，陆续增建思颜堂、思孟堂、罗氏图书馆等多座重要建筑，形成以大草坪为中心的建筑群。

其中，1894 年始建的怀施堂为标志性建筑，因纪念创始人施约瑟得名（1951 年改名为韬奋楼）。怀施堂由通和洋行设计，平面呈日字形围合布置，两进院落。西式建筑的体量之上加以中国式屋顶，四角翘起，学院希望新校舍“务将中国房屋之特质保存”“屋顶之四角，皆作曲线形”。之后陆续新建的建筑，如范文照设计的交谊楼等，大多延续了怀施堂的手法，使校园呈现出统一的中国古典复兴风格。

St. John's University, the earliest-established Christian school that has operated longer than any other university of its kind in China, is one of the most famous universities in modern Shanghai and in China at large. In 2019 it was announced as a Major Historical and Cultural Site Protected at the National Level. Today, on the campus of East China University of Political Science and Law at Edna Villas, Changning District, these buildings, combining Western and Chinese elements, look serene and imposing, telling people the history of the "Oriental Harvard".

In 1878, American Episcopalian Bishop in Shanghai, Samuel Isaac Joseph Schereschewsky, purchased 84 mu of land in the Fanhuangdu area in west Shanghai. He merged the original Baird Hall and Duane Hall and founded St. John's College the following year. In 1892, the university curriculum was set up, and the modern Western academic system (such as elective system, examination honorary system) and campus activities were introduced in an all-round manner.

Thanks to the efforts led by the principal Francis Lister Hawks Pott, the college turned into the Episcopalian private university in 1902 and was renamed St. John's University. In 1905, it was registered as a comprehensive missionary university in Washington, D.C., with the five colleges of arts, science, medicine, engineering and theology, as well as affiliated preparatory schools. Many famous universities such as Harvard University and Yale University all recognized the academic qualifications issued by the university. Its graduates could go directly to the Graduate School of American universities. It also became the first university in China to award a bachelor's degree.

After the August 13 Incident broke out, St. John's University moved to the Shanghai International Settlement and formed Shanghai United Christian University with University of Shanghai, Soochow University and Hangchow University. It moved back to its original site in 1939.

In 1950, it broke away from the Episcopal Church and became an independently Chinese-run school. In 1952, the university saw its colleges and departments reconstructed, with the disciplines incorporated into relevant universities in Shanghai, and East China University of Political Science and Law was established on its original site.

In the 19th century, St. John's University had only one chapel, the Schereschewsky Hall and the science hall. Amid continuous expansion, Jessfield, Melrose Hall, Seth Low Library and other important buildings were built, forming a building complex surrounding the lawn.

Among them, the Schereschewsky Hall, construction for which began in 1894, is a landmark building. It was named in memory of the founder Bishop Schereschewsky (In 1951, it was renamed Taofen Building). The Schereschewsky Hall was designed by Atkinson & Dallas Architects and Civil Engineers Ltd. It features a rectangle plane with a two-entry courtyard. The Chinese roof was added to the Western-style building, with the four corners raised. The college hoped that the new school building would "preserve the characteristics of Chinese houses" and that "the four corners of the roof will be made into curves". The follow-up buildings, such as the Jiaoyi Building designed by Fan Wenzhao, mostly continued the Schereschewsky Hall style, which allowed the campus to present a unified Chinese classical style.

# 沪江大学近代建筑

Historic Buildings of University of Shanghai

杨浦区军工路516号
516 Jungong Road, Yangpu District

“对于 1926 年来访中国的人而言，如果旅行者从上海进入这个国家，在郊外，他的轮船就会经过一所学校的建筑群，他会被告知这是由美国浸礼会办的沪江大学。”这里所描述的便是与圣约翰大学齐名，近代中国规模最大的教会大学之一——沪江大学。

1906 年，美国浸礼会在四川路创办浸会神学院，随后利用校长柏高德在军工路买下的土地，修建新校园。1907 年学生宿舍北堂建成，1909 年建成了第一栋综合性建筑思晏堂，随后以其为中心，建造若干栋建筑，形成了早期的校园布局。

1909 年浸会大学堂开设，两年后与神学院合并为上海浸会大学。1914 年中文校名定为沪江大学，1917 年在美国弗吉尼亚州注册立案，并获得学士、硕士学位授予权。

1918 年校园南扩，陆续增建了图书馆等建筑，至 1928 年已有 30 余栋，形成文、理、商、教育四个学院，沪江大学的格局基本形成。

沪江大学的校园建筑，没有像当时其他高校一样，尽力通过本土化让国人产生心理认同，而是营造了美国哥特复兴风格的异域风情，使其在近代中国大学中独树一帜。清水红砖砌筑、陡直的两坡红瓦屋面、砖混结构，以及随处可见的老虎窗、尖券双窗、小尖塔、木制垂花吊顶、玫瑰花窗等，都呈现出浓郁的美式风情。校园布局以美国弗吉尼亚州校园为范本，采用美国大学的规划设计手法，重视建筑与自然的融合，犹如一个大公园，中心景观面向黄浦江敞开。

1952 年全国高校院系调整，沪江大学撤并入各校，原址组建上海工业学校，历经多次变迁，1996 年重组为上海理工大学。2019 年，被公布为全国重点文物保护单位。

如今漫步在校园内，依然能感受到当年徐志摩在此求学时，青年学子意气风发，漫步黄浦江边的惬意时光。

"For anyone who arrived in China in 1926, if they entered the country through Shanghai, their ship would pass by a college building complex in the suburbs, and they would be told that it was the University of Shanghai run by the American Baptist Churches USA (ABCUSA)." The said University of Shanghai was once one of the largest missionary universities in modern China, as famous as St. John's University.

In 1906, the ABCUSA founded the Shanghai Baptist Theological Seminary on Szechuen. Road. Soon afterwards, a new campus was built on the land by Jungong Road, bought by their president John Thomas Proctor. The north wing of the student dormitory was first built in 1907, and the first building complex, Yates Hall,in 1909. Centering on the latter, several other buildings were erected, forming the early layout of the university.

In 1909, Shanghai Baptist College was founded. Two years later, it was merged with Shanghai Baptist Theological Seminary to form Shanghai Baptist College and Theological Seminary. It was officially renamed the University of Shanghai in 1914 and registered in Virginia, USA, in 1917, in a position to confer bachelor's and master's degrees.

In 1918, the campus began to expand southwards and a group of new buildings, including the library, were erected. By 1928, the university had more than 30 buildings in total and four faculties, covering arts, science, of business, and education.

Unlike other colleges built at the time which tended to adopt local architectural philosophies to win recognition from the Chinese people, the University of Shanghai had buildings featuring the exotic Gothic revival style typical of American buildings, which made the university architecturally unique among universities in modern China. American aesthetics are evident in the steep and double-sloped roofs covered with fine red bricks, the brick masonry structures and the dormer windows, double windows with pointed arches, small spires, suspended ceilings with wooden tassels and rose windows that can be seen across the university. Modeled on the layout of Virginian colleges and following the planning and design techniques of American universities that value harmony between man-made buildings and nature, the campus is like a large park whose central scenery opens to the Huangpu River.

In 1952, during a nationwide college system restructuring campaign, the University of Shanghai merged with several other universities and Shanghai Industry School was set up on its campus. The latter, through changes over the years, was restructured into University of Shanghai for Science and Technology in 1996. In 2019, it was announced as a Major Historical and Cultural Site Protected at the National Level.

Today, if you walk around in the campus, you can still feel the pleasure that Xu Zhimo must have felt when he walked by the Huangpu River as a promising young student at the university.

## 徐家汇天主堂

Zi-ka-wei Cathedral

徐家汇天主堂，是上海市区高度最高、规模最大的教堂，也是现存唯一的双钟塔式哥特风格教堂，为上海近代模仿西式教堂最杰出的案例，曾经是中国天主教地位最重要的主教堂。

徐家汇老堂始建于 1847 年，为带有江南传统民居特征的希腊式风格，是上海开埠后市区内建的第一座天主教堂。之后的半个世纪中，徐家汇老堂经历了多次的改扩建。1906 年开始在老堂西南侧重建的新堂，由英国著名建筑师道达尔设计，采用简化的哥特式双钟塔风格，拉丁十字式平面，五廊形巴西利卡，砖木结构。其两座相对的钟塔，高达 56 米，为上海近代教堂的至高点。

室内大厅被两列束柱分为中厅和侧廊，呈现出典型的哥特式教堂特征。中厅高敞，通高三层，两侧廊较低，高二层，中高侧窗采光，顶部均为四分尖券肋骨拱顶，后墙悬挂巨幅宗教画《最后的晚餐》。堂内楹柱 64 根，每根楹柱由 10 根小圆柱拼合，均由苏州产的金山石精刻细凿而成。这种西式柱式的中国做法（由小柱拼成大束柱），从米兰大教堂改制而来，成为之后上海教堂模仿的范式。

1910 年建成的徐家汇天主堂，以其规模宏大、造型美观、装饰华丽、工艺精湛，被誉为“中国教堂之巨擘”“远东第一大教堂”，更有 20 世纪“上海第一建筑”的美名，在 20 世纪 20 年代上海兴建高层建筑之前，徐家汇天主堂一直是市区最高的建筑物。

2013 年，徐家汇天主堂被公布为全国重点文物保护单位。今天，这里成为上海重要的旅游景点。

徐汇区蒲西路158号
158 Puxi Road, Xuhui District

Zi-ka-wei Cathedral is the highest and largest church in downtown Shanghai, and the only existing Gothic church with two bell towers. As the most outstanding example of Western-style churches in modern Shanghai, it was once the most prominent Catholic cathedral in China.

Built in 1847, the Greek-style Old Zi-ka-wei Church incorporated the characteristics of traditional houses in the Jiang'nan region. It was the first Catholic church built in downtown Shanghai after the city opened as a trading port. In the following half century, the Old Zi-ka-wei Church underwent multiple renovations and expansions, which was no different to other churches. None of the other churches in Shanghai in the 19th century, due to constraints in technology and funding, were built once and for all according to the original designs. Instead, they were constantly expanded and renovated to improve the original designs.

Zi-ka-wei Church, which sat to the southwest of the Old Zi-ka-wei Church, was designed by famous English architect William Doyle and broke ground in 1906. Mr. Doyle adopted a simplified Gothic style with two bell towers, the Latin cross plane, the five-corridor Basilica, and a brick and wood structure. Its two 56-metre-high bell towers, which stand opposite each other, stand taller than other modern churches in Shanghai.

The church is in the typical Gothic style, with its interior hall being divided into the nave and side corridors by two rows of compound piers. The spacious nave is three-storeys high and the lower side corridors are two-storeys high. There are upper windows on both sides. The top is decorated with rib vaults. On the back wall hangs a huge replica of the religious painting The Last Supper. There are 64 compound piers in the hall, each consisting of 10 small columns, all carved out of Jinshan stone from Suzhou. This combination of Chinese craftmanship and western-style columns (small columns are stacked to make a compound pier) is an adaptation of the columns in the Milan Cathedral, which set up an example for later churches built in Shanghai.

Built in 1910, Zi-ka-wei Cathedral was hailed as "the most majestic church in China" and "the grandest church in the Far East" due to its considerable size, splendid style and decoration, and exquisite workmanship. Before the completion of the HSBC building, it was been "Shanghai's No.1 building" in the 20th century. It remained the tallest building in downtown Shanghai before high-rises sprang up in the 1920s.

In 2013, it was announced as a Major Historical and Cultural Site Protected at the National Level. Now, as a part of the Xujiahui Origin under protection, it has become a major tourist attraction in Shanghai.

# 佘山天主教堂

Sheshan Basilica

1871 年始建佘山天主堂，包括一座希腊风格的山顶大堂和一座罗马风格的中山教堂，堂前设三圣亭和十多个沿山路的苦路亭。

1925 年佘山天主堂重建，最初设计了一座双钟塔式哥特风格教堂。但因工程主体须在近百米高处施工，难度巨大，没有营造厂愿意承接。后重新设计并建造。前后历时十年，成为上海近代建造时间最长、工程量最大的教堂。

教堂平面为拉丁十字形，巴西利卡式，钟楼高 38 米。大堂坐东朝西，正门朝南，正对信徒朝圣的“之”字形苦路。大堂由 40 根刻着天神和花卉图案的石柱支撑，尽端设大祭台，背后为半圆形龛，四分拱顶，设近 1000 个座位。教堂的标志性特征是其西南角的穹顶，下部建造在方形钟楼基础上，由 16 根柱子支撑着一个橄榄形圆穹顶。

佘山天主堂以文艺复兴时期的罗马风格为主，又杂糅了哥特式尖顶，希腊式柱式，西班牙风格小圆顶，中国式的清水砖墙、琉璃瓦和斗角地砖。这种建筑风格杂糅中协调，颇具海派文化的特性。教堂顶上的佘山圣母像，是圣母直立、托举双臂平展的耶稣形象，自成十字架状，为江南天主教会特别信奉。

该堂彻底摆脱了木结构体系模仿砖石结构的做法，完全用砖石建造了主体结构，即实现了由木结构体系向砖石结构体系的完整意义上的转变，因而也使其成为上海近代建造西式教堂在结构体系上最成熟的案例。

1985 年修缮后，教堂恢复使用，仍是整个远东地区最宏伟的山地朝圣教堂。1989 年被公布为上海市文物保护单位。

松江区佘山镇佘山巅
Sheshan Mountain, Sheshan Town, Songjiang District

The construction of the Sheshan Basilica began in 1871. It included a Greek-style mountaintop church and a Roman-style mid-mountain church, in front of which was a shrine to the Sacred Heart, the Virgin Mary, and St. Joseph. A dozen Stations of the Cross were constructed along the path to the church.

The church was rebuilt in 1925. The original design was a Gothic structure with two bell towers. However, as the construction needed to be completed mainly at a height of nearly 100 metres, no construction company wanted to take up the project. So it was redesigned and built. As the stones were shipped from Fujian Province to the foot of Sheshan Mountain and then carried to the top of the mountain by hand, the project progressed slowly, lasting nearly ten years. The Sheshan Basilica thus became the most time-and-effort-consuming church in modern Shanghai.

The Basilica-style church is in the shape of a Latin cross, with a 38-metre-high bell tower. The church sits in the east and faces west. The pilgrimage zigzag path leads to the main entrance which faces south. The nave is supported by 40 stone pillars carved with gods and flowers, with a large altar at the end, a semicircular niche at the back, a quadripartite vault and nearly 1,000 seats. The iconic feature of the church is the olive-shaped dome in the southwest corner, which is built on the base of a square bell tower and supported by 16 pillars.

The Sheshan Basilica is mainly in the Roman style of the Renaissance, mixed with Gothic spires, Greek columns, Israeli bell towers, Spanish domes, Chinese brick walls, glazed tiles and corner tiles. Such a combination is a quintessential feature of Shanghai culture. At the top of the church stands the statue of Our Lady of Sheshan, with Madonna standing upright and holding Jesus, whose arms are outstretched. The cross-shaped statue is worshipped particularly by the Jiangnan Catholic Church.

The church did not use a wooden structure to imitate the masonry structure. Instead, its main structure was entirely built with bricks, which represents a complete transformation from wood to masonry. Therefore, it is seen as the most mature case in the construction of Western-style churches in modern Shanghai.

The church reopened after renovation in 1985. It remains the most magnificent mountain pilgrimage church in the whole of Far East. In 1989 it was designated as a cultural site protected by the city of Shanghai.

# 圣三一基督教堂

Holy Trinity Church

黄浦区九江路201号
201 Jiujiang Road, Huangpu District

19 世纪中期以后，远眺外滩，在一片外廊式建筑背后，一座红砖砌筑的哥特式钟塔高高耸立，成为天际线上最醒目的地标和制高点，这就是上海开埠后建造的第一座新教教堂——圣三一堂。这座当时远东级别最高的教堂，目前也是上海现存最早的新教教堂。

圣三一堂又名“红礼拜堂”，属英国圣公会，1847 年始建于江西路九江路口。1866 年始建新堂，1869 年建成。1875 年，英国维多利亚女王把圣三一堂升格为安立甘北华（圣公会）教区的主教座堂，故又称“安立甘大教堂”，由坎特伯雷大主教主管。1893 年，教堂东北侧增建一座方形平面、尖锥形屋顶的哥特式钟塔（后毁于 1966 年），其形式模仿了法国夏特尔大教堂西南角的塔楼，至此才形成了完整的罗马风教堂的形制。1914 年，教堂又安装了第一个、也是远东最大的电动鼓风的大型管风琴。

在近百年的时间里，圣三一堂仅供圣公会的侨民礼拜，而非圣公会的外侨则要到天安堂礼拜。直至 1937 年闸北圣保罗堂在淞沪会战中被炸毁，中国信徒才被允许借该堂礼拜，圣三一教堂也才由一个侨民教堂转变为城市公共教堂。

教堂平面为拉丁十字形，五廊形巴西利卡式，砖石结构，高约 19 米。中厅为木拱结构，剪刀形木屋架，石铺屋顶。红砖砌筑的外墙，于发券处夹有灰砖，成斑马纹式样，独具地域特征。

座席为靠背穿藤的长椅，窗框和椅背都钉着捐献者姓氏的铭牌，一派欧洲古教堂遗风。

教堂采用带有哥特式复兴特征的晚期罗马风风格，三面皆有连续券柱敞廊，西侧为古安立甘式穹顶结构至圣所。细部设计则显露出罗马风与哥特式的混合，如入口门廊的中券为圆券，两侧小券为尖券，即同时使用尖券和圆券。哥特复兴风格的钟塔与教堂主体脱开，因地制宜地贴合了基地的形态。教堂东侧有大片绿地，草木茂盛。1966 年以后，草地改建成街心花园，与教堂隔离，成为近代早期外滩地区重要的市民活动空间。唯有两棵市区少见的百年银杏树，秋中一片金黄，与教堂红色相辉映。

20 世纪 50 年代，圣三一堂曾作为中华圣公会的主教堂，教仪仍用圣公会传统。2005 年后，经大修恢复了昔日的容貌。2019 年被公布为全国重点文物保护单位。

After the mid-19th century, if you looked beyond the Bund, you could see, behind a collection of ordinary veranda-style buildings, a red-brick Gothic-style bell tower standing out. The most striking landmark and the commanding point on the skyline was the Holy Trinity Church, the first Protestant church built after Shanghai opened as a trading port. It was the most significant church in the Far East at the time and is also the oldest existing Protestant church in Shanghai.

The Holy Trinity Church, also known as the "Red Church", is an Anglican church. It was first built in 1847 at the junction of Jiujiang Road and Jiangxi Road.The construction of the new church began in 1866 and was completed in 1869. In 1875, Queen Victoria of England elevated the Holy Trinity Church to the Anglican Cathedral in the Diocese of North China, so it was also known as the "Anglican Cathedral", headed by the Archbishop of Canterbury. In 1893, a gothic bell tower with a square plan and a tapered roof was added to the northeast side of the church (destroyed in 1966), which imitated the tower at the southwest corner of the Chartres Cathedral in France, thus completing the Roman style of the church. In 1914, the first and largest electric pipe organ in the Far East was installed in the church.

For nearly a century, the Holy Trinity Church served only Anglican foreign residents while non-Anglican foreign residents had to worship at the Church of Heavenly Peace. It was not until 1937, when St. Paul's Church in Zhabei was bombed in the Battle of Shanghai, that Chinese believers were allowed to worship here, and the church then changed from a church for foreign residents to a public church of the city.

With a Latin cruciform plane and a five-corridor Basilica, the church is a nearly 19-meter-high masonry structure. The nave is a wooden arch structure, with a scissor-shaped wooden frame and a stone roof. On the outer walls, the red bricks are mixed with gray ones at the arches, forming a distinctive zebra-striped pattern. The whole church is in the old European style due to the pews with rattan backs and the window frames that all bear nameplates of the donors' surnames.

The church is in the late Roman style with Gothic revival characteristics and continuous verandas supported by arched columns on three sides. On its west side stands the Sanctuary, which has an ancient Anglican dome. The details show a combination of Roman and Gothic styles. For example, at the entrance porch, the arch in the middle is round-shaped and the smaller arches on both sides are pointed. The bell tower in the Gothic revival style, which fits the foundation perfectly, is separated from the main body of the church. On the east side of the church lies a large piece of green land with lush vegetation. After 1966, the green land was transformed into a street garden, isolated from the church, and became a major public space in the Bund area. In autumn, two golden centuries-old ginkgo trees, rarely seen in downtown, stand against the red church.

In the 1950s, the church served as the Anglican cathedral of the Chung Hua Sheng Kung Hui. It has been overhauled since 2005. In 2019, it was announced as a Major Historical and Cultural Site Protected at the National Level.

## 摩西会堂旧址

Former Ohel Moshe Synagogue

虹口区长阳路62号
62 Changyang Road, Hongkou District

提起上海的犹太人，就不能不提虹口的提篮桥地区，这里是国内唯一能反映二战时期犹太难民生活的历史街区。这里有一栋饱受风霜的教堂，就是有“诺亚方舟”之称的摩西会堂。

摩西会堂，又称华德路会堂，始建于 1907 年，因纪念俄罗斯犹太人摩西・格林伯格而得名。之后随着上海的俄籍犹太人日渐增多，便于 1927 年迁址新建，供俄罗斯和中欧犹太人使用。

这座外廊式风格的建筑，主体三层，局部四层，平面呈凸字形，砖木结构，青砖墙面，红砖水平带状装饰，拱廊入口处有犹太教标志，中央为礼拜堂，可容纳 300 人。楼梯每一转角的下部，都有俄籍犹太裔居室中常见的葱头状木装饰。

二战期间，大批欧洲犹太难民来到上海。摩西会堂成了难民们的避难所，通过信仰维持着生活的希望，并坚持反法西斯斗争。二战后，尽管大批犹太人迁居世界各地，但对摩西会堂的感念已深藏心底，视上海为“第二故乡”。

1998 年摩西会堂更名上海犹太难民纪念馆，由旧址和两个展示厅组成，见证了那段艰难历程。2007 年全面修缮，恢复了昔日的风貌。2014 年被公布为上海市文物保护单位。

如今，这里成为许多犹太人来上海的必到之处，驻足摩西会堂旧址前，面对刻有一万多犹太人难民名字的铜墙，不禁触景生情。一段温暖的驻留，让避难不再灰暗，也让每一位游客真切感受到这座城市的包容和友善。

You cannot talk about Jews in Shanghai without mentioning the Tilanqiao area in Hongkou. Here stands the weathered Ohel Moshe Synagogue, dubbed the Noah's Ark.

First built in 1907, the Ohel Moshe Synagogue, also known as the Ward Road Church, was named after Moshe Greenberg, a member of the Russian Jewish community. As the number of Russian Jews coming to Shanghai grew, a new synagogue was built for Russian and Central European Jews in a different place in 1927.

The Synagogue is in the veranda style, with three floors in the main body and four in part. With a convex plane, the brick and wood structure has blue brick walls, which are decorated by horizontal red brick ribbons. The entrance of the arcade sports the Jewish symbol. The nave can accommodate 300 worshippers. Beneath each corner of the staircases is a wooden onion-shaped adornment, commonly seen in the homes of Russian Jews.

During World War II, a large number of European Jewish refugees came to Shanghai. The Synagogue thus became a sanctuary for the refugees, who kept their hope for life through faith and persevered in the struggle against fascism. After World War II, the Jews maintained their feelings for the Synagogue and saw Shanghai as their "second home".

In 1998, the Synagogue was renamed the Shanghai Jewish Refugees Museum, and two exhibition halls were added. In 2007, it was fully renovated and restored to its former appearance. In 2014 it was designated as a cultural site protected by the city of Shanghai.

Today, it is a must-see for many Jews visiting Shanghai. They can stand in front of the Synagogue, looking at the bronze wall with the names of more than 10,000 Jewish refugees. The Synagogue, which served as warm shelter making the difficult times less dark, lets every visitor really feel the inclusiveness and kindness of Shanghai.

# 新乐路东正教堂

Russian Orthodox Mission Church on Xinle Road

20 世纪初的上海法租界，西班牙风格建筑随处可见。但在新乐路上，却有一座小巧玲珑、造型别致的俄罗斯风格教堂。那层层拱起的屋檐和“洋葱头”式的穹顶，在周边环境中特别显眼，这就是上海现存规模最大的东正教堂——新乐路东正教堂。2014 年被公布为上海市文物保护单位。

俄侨自 19 世纪在沪定居开始，就筹划建造东正教堂。1905 年，闸北建成上海教区的第一座东正教堂。1932 年一・二八事变中，俄侨的主教堂闸北主显堂被炸毁，加之俄侨大多迁居法租界，急需在租界内新建教堂。圣尼古拉斯教堂和新乐路教堂同年修建、同年竣工，也是今天上海仅存的两处东正教教堂。由于资金有限，最初建造了一个木结构的临时教堂——天使长加夫里洛教堂。1936 年建成新堂。作为东正教在沪的主教座堂，仿莫斯科基督救世主大教堂设计，为典型的俄罗斯式拜占庭风格，集中式布局，希腊十字平面。中部为孔雀蓝色的主穹顶，高达 35 米，状似洋葱，四角均有小穹顶。这些洋葱式的大小穹顶，和奶白色的墙面，在蓝天白云下浑然一体，呈现出浓郁的俄罗斯风情。

2007 年教堂修复时，推断主穹顶的白色涂料下极有可能藏有历史原物的湿壁画。后将多层涂料清除后，湮没了近半个世纪的九幅珍贵壁画重见天日。

In the early 20th century, Spanish-style structures were seen everywhere in the French concession of Shanghai. Yet on Xinle Road, there was a small, chic, Russian-style church, whose arched eaves and onion-shaped domes stood out from its neighbouring buildings. That is the largest existing Orthodox church in Shanghai—the Russian Orthodox Mission Church. In 2014 it was designated as a cultural site protected by the city of Shanghai.

After they settled in Shanghai in the 19th century, the Russians were planning to build Orthodox churches in the city. In 1905, Shanghai Eparchy's first Orthodox church was built in Zhabei District. In the January 28 Incident of 1932, the church in Zhabei was bombed, and as most of the Russians moved to the French concession, a new church was needed there. The St. Nicholas Church and the Church on Xinle Road were established in the same year; these two churches are the only remaining Orthodox churches in Shanghai today. At first, a temporary wooden structure, the Archangel Gavrilo Church, was built for a lack of money. A new church was completed in 1936. A cathedral of the Orthodox Church in Shanghai, the church on Xinle Road was designed in imitation of the Cathedral of Christ the Saviour in Moscow. It is in the typical Russian Byzantine style, with a centralized layout and a Greek cross plane. In the middle there is the 35-metre-high onion-shaped peacock blue main dome, with small domes on four corners. The picture of the onion-shaped domes and the creamy-white walls presents a strong Russian flavour in good weather.

When the restoration started in 2007, it was inferred that the original frescoes was under the white paint of the main dome. After the multiple layers of paint were removed, nine precious frescoes were rescued after being buried for nearly half a century.

# 都会映象

Metropolis Impression

# “大上海计划”公共建筑群

The Public Buildings of the “Greater Shanghai Plan”

杨浦区江湾五角场
Jiangwan-Wujiaochang, Yangpu District

在江湾五角场的东面，分布有多处颇具特色的建筑，包括旧上海特别市政府大楼、江湾体育场、旧上海市图书馆、旧上海市博物馆、旧上海市立医院、上海市立卫生研究所等，颇具规模和特色。

1927 年 7 月，国民政府宣布建立“上海特别市”，英文名“City of Greater Shanghai”，直接隶属行政院，即“中央政府直辖市”，1930 年 7 月起，改称“上海市”，英文名称照旧。

上海特别市政府则设在远离市区的“市政府路”（今平江路），1930 年拟订“大上海计划”，拟在市区东北的江湾，东临黄浦江的 7000 余亩土地上建造一个“新上海”，将市政府及下辖主要机构迁移过来，并逐渐把上海的市中心区向此转移，形成一个上海“新城”。

一个庞大的城市规划和建设，必然会涉及建筑的样式、风格、色彩、高度，这对市政府大楼尤其重要。“大上海”的市政府、博物馆、图书馆等主要建筑的样式被确定为仿中国古典建筑。市政府的建筑体量很大，像北京的故宫建筑，后来决定在色彩上动脑筋，将传统宫殿的黄琉璃瓦改为青瓦（实际是深绿），墙体白色，加上红色的柱子。而位于市政府大厦两侧的博物馆、图书馆采用了中国建筑中的钟楼和鼓楼样式。

市政府大厦于 1931 年 6 月奠基，开工不久一・二八淞沪抗战爆发，工程被迫拖延，一直到 1933 年 9 月竣工，市政府分批迁入新址。当时实际完成的建筑还有上海市体育场、上海市立医院、国立音专、上海市广播电台及材料研究所、中国航空协会飞机楼等。1937 年抗日战争全面爆发，部分建筑毁于战火，上海市政府内迁重庆，这个规模空前的“大上海计划”就此夭折。

如今，原部分区域被公布为“历史文化风貌区”，留下的部分建筑也被公布为上海市文物保护单位，如上海体育学院院部办公楼、江湾体育场、杨浦区图书馆新馆等。

江湾体育场
Jiangwan Stadium

①
②
③

① 上海体育学院
Shanghai University of Sports

② 长海医院飞机楼
Changhai Hospital Aircraft Building

③ 杨浦区图书馆
The Yangpu District Library

To the east of Jiangwan-Wujiaochang lies a cluster of magnificent buildings that resemble Chinese palaces, including the former City Hall of Shanghai Special City Government, Jiangwan Stadium, the former Shanghai Municipal Library, the Old Shanghai Municipal Museum, the Old Shanghai Municipal Hospital, the Shanghai Municipal Health Research Institute, and so on.

In July, 1927, the National government announced the establishment of the"City of Greater Shanghai", a "municipality directly under the National Government" directly under the Executive Yuan. In July, 1930, it was renamed "Shanghai City" in Chinese, whereas its English name remained unchanged.

The Government of Shanghai Special Administrative City was located on the "City Government Road" (now known as Pingjiang Road), far from the downtown area. In 1930, the Government of Shanghai Special Administrative City approved the Greater Shanghai Plan and allocated more than 7,000 mu for the construction of a "new Shanghai" to the west of the Huangpu River in Jiangwan, northeast Shanghai, planning to move the government headquarters and its major organs there. By doing so, the government aimed to gradually move the downtown area of Shanghai to this place, giving rise to a "new city of Shanghai".

Architectural style, characteristics, colors and height would surely be dealt with in the planning and construction of such a colossal project, especially when it came to the city hall. Eventually, it was decided that the major buildings, including the city hall, museum and library, would follow the style of traditional Chinese architecture. As the city hall would be a building of almost the same magnitude as that of the Forbidden City in Beijing, it was decided to make a few color changes. So, it would be covered with grey tiles (which then became dark green), instead of yellow glazed tiles as in the case of traditional palace buildings, and the walls would be white, coupled with red columns. The museum and the library on the two sides of the city hall would respectively be in the style of the Chinese bell tower and the Chinese drum tower.

The city hall was inaugurated in June 1931, but construction was put off soon afterwards because of the outbreak of the January 28 Incident. It was not finished until September 1933, and the government organs moved into the new building successively. Other structures completed at the time include Shanghai Municipal Stadium, Shanghai Municipal Hospital, the National Conservatory of Music, Shanghai Radio Station, Materials Research Institute and the fuselage-shaped building for the China Civil Aviation Association. After the full outbreak of the War of Resistance against Japan in 1937, some were destroyed in the flames of war. The Shanghai City Government then moved to Chongqing. The "Greater Shanghai Plan" unprecedented in scale, came to a premature end.

Today, part of the area has been announced as a historic and cultural conservation area and some of the preserved buildings were declared as protected monuments of Shanghai, such as Shanghai University of Sports Office Building, Jiangwan Stadium, the Yangpu District Library and so on.

# 大世界游乐场

The Great World

黄浦区西藏南路1号
1 South Xizang Road, Huangpu District

大世界坐落在延安东路与西藏南路交叉路口，原来全名为“大世界游艺场”，是旧上海著名的游乐场，其建筑也是上海标志性建筑。

1917 年，黄楚九集资 80 余万元组织“大发公司”，觅得法租界敏体尼荫路与爱多亚路（西藏南路与延安东路）东南转角的一块九亩八分的土地，建成一幢面积达 14700 平方米的三层砖木结构建筑。他又邀请社会名流给多处景点取名题词，确定了“飞阁流丹”“层楼远眺”“亭台秋爽”等十大景点，经过报纸的渲染，这个还没开业的游艺场就带给市民一探究竟的强烈冲动。

大世界的营业时间从下午 1 时至午夜 12 时，节目丰富，屋顶辟有花园，设有茶室，还配了几台固定在架子上的望远镜，可以观看数里之外的上海全景。

大世界生意旺盛，游客众多，三层砖木结构似乎难以承受如此的践踏。1924 年在原址重建钢筋水泥的多层建筑，建筑的主立面设计在敏体尼荫路与爱多亚路交岔口，主建筑为四层，在四层以上设计为由 48 根圆柱支撑的六面四层尖塔顶，使建筑高度达到 55.3 米，也是当时上海的至高点，在远处就能看到这造型别致的塔顶，游客购票后也可以升到塔顶俯瞰上海。

主立面的两翼均为与之相通的四层建筑，形成一个扇形，扇形背面则建有露天舞台，可以表演大型的杂技魔术，建筑之间用天桥相连，既是楼与楼之间的通道，又是露天舞台的看台，每逢演出，人们一定会占据这里，从高处更清楚地观看。相当长的时间里，一直有“没去过大世界，等于没有到过大上海”之说。

1954 年 7 月，上海市文化局接管了这里，一度改称人民游乐场，1956 年仍恢复使用“大世界”之名。1974 年改名上海市青年宫，由共青团上海市委管理。1987 年作为“大世界游乐中心”开业。1989 年被公布为上海市文物保护单位。如今，大世界作为上海非物质文化遗产展示馆对外开放，由黄浦区政府管理。

Located at the intersection of East Yan'an Road and South Xizang Road and formerly known as "Great World Amusement Park," the Great World is a famous amusement park in old Shanghai, and its building is also a landmark in the city.

In 1917, Huang Chujiu raised more than 800,000 yuan to organize the"Dafa Company" and purchased a plot of 9.8 mu in the southeast corner of Boulevard de Montigny and Avenue Edward VII (now known as South Xizang Road and East Yan'an Road) in the French Concession. It was a three-storey brick and wood structure covering an area of 14,700 square meters. He invited celebrities to visit and give advice, and asked them to come up with names and inscriptions for the attractions. Eventually, ten major attractions were confirmed, such as "Red Suspending Pavilion" "Overlooking Tower" and "Terrance for Autumn Scene" and so on. With a media hype, the amusement park aroused an immense curiosity of the public even before its opening.

The business hours of Great World were from 1:00 p.m. to 12:00 p.m. and there were variety of shows. There was a garden on the roof, facilitated with a tea room and a couple of telescopes fixed on stands by which one could enjoy a panoramic view of Shanghai several miles away.

The Great World saw thriving business and streams of visitors. The three-storeyed brick-wood structure was unable to bear such heavy load. In 1924, a reinforced concrete multi-storey building was rebuilt on the site. The main façade of the building was designed to face the intersection of Boulevard de Montigny and Avenue Edward VII. The main building has four storeys and above the fourth floor is a six sided four-storey tower supported by 48 columns, making it the highest building of Shanghai at that time with a height of 55.3 meters. The unique tower peak could be seen from a distance and tourists would go to the top to overlook Shanghai after buying tickets.

The two wings are four-storey buildings connected to the main facade, forming a fan-shaped structure. On the back of the fan-shaped structure is an open-air stage, where large-scale acrobatic and magic shows could be staged. The buildings are connected by flyovers, serving both as the passageways between buildings as well as grandstands of the open-air stage. People would take a place here to have a better view of the staged performances. For quite a long time, there had been a saying that,"Tell me not you've been to Great Shanghai if you haven't been to the Great World."

In July 1954, it was taken over by the Shanghai Municipal Administration of Culture and was once renamed the People's Amusement Park. In 1956, it restored its name as "Great World." In 1974, it was renamed Shanghai Youth's Palace and was managed by the Shanghai Municipal Committee of the Communist Youth League. In 1987, it reopened in the name of "Great World Amusement Centre." In 1989, it was designated as a cultural site protected by the city of Shanghai. Today, the Great World opens to the public as the Shanghai Intangible Cultural Heritage Exhibition Hall, administered by Huangpu District government.

# 上海音乐厅
Shanghai Concert Hall

黄浦区延安东路 523 号，原南京大戏院
523 East Yan'an Road, Huangpu District, Former Nanking Theatre

在上海的优秀近代历史建筑中，像上海音乐厅这样被整体平移过的数量极少。

作为上海音乐厅前身的南京大戏院，是上海在电影自 20 世纪 20 年代末期进入有声时代后兴建的豪华影院，它在建成时就引进了西电公司的维泰风和 Fox 公司的慕维通两种有声片放映设备，同时，还斥资进口安装了空气调节系统，最早在沪上影院中真正做到了“冬暖夏凉”。

在建筑设计上，南怡怡公司的创立者何挺然租下潮州会馆的地皮，就请来了有美国宾夕法尼亚大学留学背景的中国建筑师——范文照和赵深担任设计。影院北立面雨篷上方用爱奥尼亚柱式形成的三联壁龛和圆拱窗，楣梁上檐口下由李金发创作的浮雕，门厅可见仿巴黎歌剧院的大楼梯，休息大厅的 16 根褐色圆柱，二层的华丽罗马柱式组成的廊道，以及柱头和柱身的黑白大理石色彩对比，种种细节都呈现出浓重的西方古典主义建筑风格。

南京大戏院开业后获得了联美、米高梅、二十世纪福斯（今译“福克斯”）等多家好莱坞公司影片的首轮放映权，此外也常有音乐歌舞及戏剧演出。有趣的是，它有着迥别于一般电影院的声学特点：混响时长超过了有声片放映需要的 0.8 秒，达到 1.5 秒左右，这一点也为它日后成为专业音乐厅奠定了基础。

1950 年 11 月 18 日，南京大戏院更名北京电影院。1959 年，为庆祝中华人民共和国成立十周年和筹备“上海之春”音乐舞蹈节，南京大戏院又以它极为出色的声学效果而被选作演出场所，并改名为“上海音乐厅”，成了几代上海人音乐记忆中最为重要的地标。1989 年被公布为上海市文物保护单位。

2002—2004 年，上海延安东路“八仙桥”地块的改造，让这栋老建筑得以旧貌换新颜，经过精密测算、筹划和紧张施工，它被整体抬升 3.38 米、平移至距原址东南 66.46 米的新处所，随后的精心修缮，既保护了这座欧式古典建筑的历史原真性，又增加了建筑面积，完善了功能和设施。2019 年 3 月起，音乐厅再次修缮，于 2020 年 9 月 6 日重新开业向公众开放。

Very few of the modern Heritage Architecture in Shanghai have been moved like the Shanghai Concert Hall.

The Nanking Theatre, the Shanghai Concert Hall's predecessor, was a luxury cinema built in Shanghai in the late 1920s with the advent of sound films. At the very beginning, it introduced two kinds of sound film projection equipment, namely Vitaphone of Western Electric and Movietone of Fox Film Corporation Fox Film Corporation. At the same time, it invested heavily in an air conditioning system, becoming the first cinema that was "warm in winter and cool in summer" in Shanghai.

In terms of architectural design, He Tingran, the founder of Nanyiyi Company, rented a plot of Chaozhou Guild Hall, and then invited Chinese architects, Fan Wenzhao and Zhao Shen, who had studied at University of Pennsylvania in the United States, to design the building. There are triple niches and circular arched windows formed by Ionic orders above the canopy on the north façade of the cinema, and reliefs created by Li Jinfa below the eaves and above the lintel. The staircase in the foyer is in imitation of the Palais Garnier (Paris Opera), and there are 16 brown columns in the lounge. Along the corridor on the second floor stand magnificent Roman orders, whose black capitals and white shafts make a striking contrast. The Western classical architectural style is reflected in every detail.

After its opening, Nanking Theatre obtained the first-round screening rights of United Artists, MGM, 20th Century Fox and many other Hollywood companies. In addition, it often presented music, dance and drama performances. Interestingly, it had acoustic characteristics quite different from that of ordinary cinemas: its reverberation time was longer than the 0.8 second required for a sound film to be shown, reaching about 1.5 seconds, which also laid the groundwork for its future development into a professional concert hall.

On November 18, 1950, the Nanking Theatre changed its name to Beijing Cinema. In 1959, in order to celebrate the 10th anniversary of the founding of the People's Republic of China and prepare for the "Shanghai Spring International Music Festival", the Nanking Theatre was selected as the performance venue for its excellent acoustic effect and was renamed as the "Shanghai Concert Hall", which became the most important landmark in the music memory of several generations of Shanghai people. In 1989, it was designated as a cultural site protected by the city of Shanghai.

Thanks to the renovation of the "Baxianqiao" plot on East Yan'an Road in Shanghai from 2002 to 2004, the old building took on a brand-new look. After precise calculation, planning and intensive construction, the whole building was lifted up by 3.38 metres and transported to the new place 66.46 metres southeast of the original site. The meticulous renovation afterwards not only has preserved the historical authenticity of the European classical building, but also has increased the floor area and improved its functions and facilities. From March 2019, the Shanghai Concert Hall was closed again for refurbishment. It reopened to the public on September 6, 2020.

# 大光明电影院
Grand Theatre

黄浦区南京西路 216 号
216 West Nanjing Road, Huangpu District

坐落在人民广场文化圈内的大光明电影院，迄今近 90 年历史，建成时曾被誉为“远东第一电影院”。

原址的老大光明影戏院于 1928 年闪亮面世。1932 年，英籍广东人卢根与美国商人组成联合电影公司，租赁大光明及其附近房产，请邬达克设计重建，并更名为“大光明大戏院”，1933 年 6 月 14 日首映好莱坞电影《热血雄心》，开始其辉煌时代。1949 年以前，大光明大戏院主要放映美国福克斯、米高梅等公司的原版片，也是工部局音乐会的常驻地，还是第一家使用译意风（类似同声翻译，Earphone）的影院，堪称引领摩登生活的时尚标杆。

大光明大戏院实际上是一座集影院、舞厅、咖啡馆、弹子房等于一身的娱乐综合体。它位于错综复杂的旧建筑夹缝中，沿静安寺路（今南京西路）门面不宽，大部分是需要保留的店面，凤阳路更是只有一条狭长的逃生通道。西面紧贴里弄，东侧与派克路（今黄河路）之间夹着卡尔登大戏院（已拆除）——一座邬达克在克利洋行时期设计的古典风格的影院。

在邬达克的上海实践中，大光明大戏院的历史背景和基地条件都最为复杂，但也最体现其设计功力。因为基地狭长且不规则，设计几易其稿才最终确定。观众厅平行基地长边布置成钟形，与门厅轴线有 30 度扭转。大厅上下两层近 2000 个软座，容量当时居全国之首，内部采用暗槽灯照明，喷射式冷气。两层休息厅设计成腰果形，与流线型的门厅浑然一体。两部大楼梯直通二楼，休息厅中央还建有灯光喷水池。

建筑外观是典型的现代装饰艺术风格，立面上横竖线条与体块交错。入口乳白色玻璃雨篷上方是大片金色玻璃，还有一个高达 30.5 米的长方形半透明玻璃灯柱，夜晚尤为光彩夺目。

1949 年后，大光明大戏院更名为大光明电影院。1989 年被公布为上海市文物保护单位。历经数次改造，最后一次整修在尽可能恢复历史原貌的同时，更新了设施，新增了五个小厅、屋顶花园和餐厅，于 2009 年 1 月 19 日重新开业。

The Grand Cinema, located in the cultural hub of the People's Square, has a history of nearly 90 years. It was widely known as "the best cinema in the Far East" upon its completion back then.

In 1928, the old Grand Cinema & Theatre made its brilliant debut. In 1932, Lo Kan, a British Cantonese, established United Theatres Inc. with an American businessman. They rented the Grand Cinema and its nearby properties. He invited Ladislav Hudec to redesign the cinema and gave it a new name of the "Grand Theatre." On June 14, 1933, it premiered the Hollywood film *Hell Below*, embarking on the road to glory. Before 1949, the Grand Theatre mainly screened films in English from Fox, MGM and other American companies. It was the regular spot for the concerts orga-nized by the Municipal Council. It was also the first cinema to introduce Earphones (similar to simultaneous interpretation), and was regarded as a bellwether of fashion-able and modern trends.

The Grand Theatre was actually an entertainment complex with cinemas, dance halls, cafés, and billiard rooms. It flourished in a cramped space surrounded by old buildings: a narrow façade along Bubbling Well Road (West Nanjing Road), where most of the storefronts were to be reserved; and Fengyang Road was nothing but a narrow escape route. It was a lane that lies in the west. The Carlton Theatre (a classical style cinema designed by Hudec while he was working for Curry R.A, now demolished) was sandwiched between it and Park Road (Huanghe Road) in the east.

Among Hudec buildings in Shanghai, the Grand Theatre is the most challenging one due to its extremely complicated historical background and foundation conditions. However, it is also an epitome of his amazing talent in architecture. As the foundation was narrow and irregular, Hudec modified his drafts time and again before the design was finalized. The long side of the parallel foundation of the auditorium was arranged in a bell shape, forming a 30 degree to the axis of the foyer. There were nearly 2,000 soft seats on the upstairs and downstairs of the hall, boasting the largest capacity in China at that time. The interior of the hall adopted cove lighting and jet air-conditioning. The two-storeyed lounge was designed in the shape of cashew to harmonize with the streamlined foyer. The two large stairs led directly to the second floor, and there was a light fountain in the centre of the lounge.

The architectural exterior was typical of modern Art Deco style, with vertical and horizontal lines and blocks interlaced on the façade. On top of the opal glass cano-py was a large piece of golden glass and a 30.5-meter-high rectangular translucent glass lamp post, which shone with dazzling brilliance at night.

After 1949, the Grand Theatre was once renamed the Grand Cinema and went through several renovations.In 1989, it was designated as a cultural site protected by the city of Shanghai. In the lat-est renovation, the facilities were updated while its historic appearance was restored as much as possible. Moreover, five small studios, a roof garden and a restaurant were added. It was reopened on January 19, 2009.

## 美琪大戏院

Majestic Theatre

美琪大戏院，英文名 Majestic Theatre，位于静安区江宁路 66 号，建成于 1941 年，建筑师范文照设计，原为放映好莱坞电影的著名首轮影院，现由上海文广演艺（集团）有限公司管理运营，用作综合性的影剧演出场所，也是“上海国际电影节”“上海国际艺术节”的定点演映场馆。1989 年列入上海市文物保护单位名单。

这座位于静安区最繁华地段——南京西路商圈的影剧院，见证了上海现代影剧业的发展历程。

1941 年 10 月 15 日，美琪大戏院揭幕开业，由资深电影经营者何挺然的亚洲影院公司投资建造，公司出版的《亚洲影刊》为已确定的英文名称向影迷们征集中文译法，在收到的两千多封来信中，最后选用了七位观众提议的“美琪”，在谐音之外，更取意“美轮美奂，琪玉无瑕”。

美琪大戏院在风格上是上海后期装饰艺术派建筑的代表，这一始自 20 世纪 20 年代末的新建筑主流，构成了上海城市形象的重要传统。

影院拥有 1600 多个座位，开幕伊始，它便与同属于亚洲影院公司的大光明、国泰和南京大戏院一道，成为多家好莱坞公司出品影片的首轮放映影院。

抗战胜利后，除继续作为首轮影院放映了大量好莱坞影片

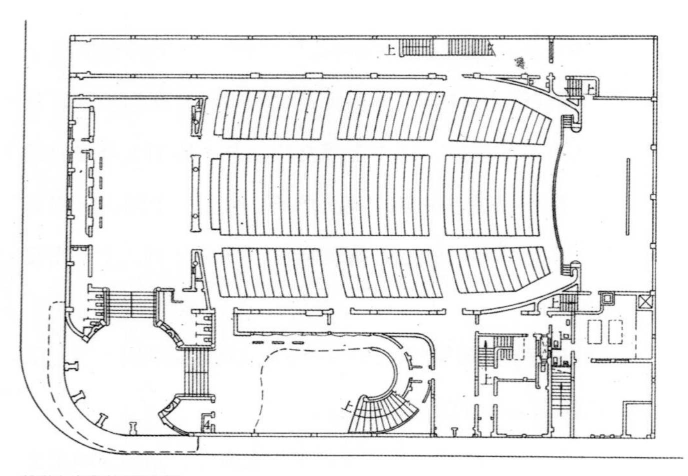


美琪大戏院底层平面图
Ground floor plan of the Majestic Theatre

20世纪50年代的戏院外景
Exterior scene of the Majestic Theatre in the 1950s

外，抗战期间“蓄须明志”的梅兰芳曾于 1945 年 12 月末起在“美琪”首次登台演出，一共十天，一时轰动上海。1949 年以后，“美琪”继续发挥着上海重要文化场馆的作用，还加强了演剧功能。

“美琪”曾多次改扩建，2016 年初，为时五年的修缮工程竣工，本着“修旧如故”的原则，努力恢复美琪大戏院海上影剧之宫的风貌。

The Majestic Theatre, located at 66 Jiangning Road, Jing'an District, was built in 1941 with the design of architect Robert Fan Wenzhao. It used to be a famous cinema for screening first-run Hollywood films. Now it is managed and operated by the Shanghai Media Group (SMG) as a comprehensive venue for screening films and staging plays. It is also a regular performance & screening venue for the "Shanghai International Film Festival" and the "Shanghai International Art Festival". In 1989, it was listed in the protected monuments of Shanghai.

Located in the most prosperous area of Jing'an District, the West Nanjing Road Business District, the theatre has witnessed the development of Shanghai's modern film and stage industry.

On October 15, 1941, the Majestic Theatre unveiled and was opened for business. The cinema was investment from the Asia Film Company owned by He Tingran, a veteran film operator. Asia Film magazine published by the company called for Chinese translations from fans for its confirmed English name. Among more than 2,000 letters received, " 美琪 " proposed by seven people was chosen. It sounded like "majestic", and also implied the meaning of "beautiful and magnificent, a flawless jade".

In terms of style, the Majestic Theatre is representative of Shanghai's later Art Deco school. This new architectural mainstream, which started from the end of 1920s, played an important role in Shanghai's urban image.

With more than 1,600 seats, the Majestic Theatre, together with the Grand Theatre of Asia Film Company, Cathay Theatre and Nanking Theatre, became the first run cinemas for films produced by Hollywood companies.

After the victory of the War of Resistance, the Majestic Theatre continued to screen a large number of Hollywood films as the first run cinema. It also arranged the debut of Mei Lanfang who kept his beard to show his high ideals during the War of Resistance for ten days starting from December 1945, which caused a sensation in Shanghai. After 1949, the theatre continued to play an important role as an important cultural venue in Shanghai and strengthened its stage performance function.

The Majestic Theatre has been rebuilt and expanded many times, and the latest five-year renovation was completed in early 2016. In line with the principle of "restoring the old as the old", the design and construction teams made every effort to restore the styles and features of the Majestic Theatre.

# 国际饭店

Park Hotel

据说，许多人都曾因仰视它而把帽子掉在了地上，这个传说佐证了国际饭店多年享有的“远东第一高楼”美誉，它保持的上海建筑高度的神话长达半个多世纪，楼顶旗杆的中心位置还被定义为上海城市测绘的零坐标。

国际饭店，亦称上海四行储蓄会，24 层，地上 22 层，地下两层，总高 83.8 米。至 1983 年上海宾馆建成之前，在 53 年里保持着上海和全国最高建筑纪录。1930 年，由金城、中南、大陆和盐业四家银行联合组成的“四行储蓄会”看到地价飞涨，房地产业利润丰厚时，决定投资高层现代旅馆，选址在面对跑马厅的静安寺路（今南京西路）派克路（今黄河路）转角，建筑英文名“派克饭店”正由此而来。

国际饭店无疑是邬达克建筑作品中质量最高、影响最大的一个。他最终赢得这座大厦的设计竞赛，是因为在高层建筑结构和技术上大胆突破，解决了上海软土地基沉降的致命问题。400 根 33 米长的木桩和钢筋混凝土筏式基础，上层采用质量轻、强度大的合金钢结构，这些举措使国际饭店在同时期兴建的上海高层建筑中沉降量最小。

大楼底层主要是“四行储蓄会”的营业大厅，金库设在地下室，转角才是旅馆门厅，这跟今天的状况正好相反。二层餐厅朝南是大面积出挑的落地玻璃窗，可以一览无遗地俯瞰跑马厅。建

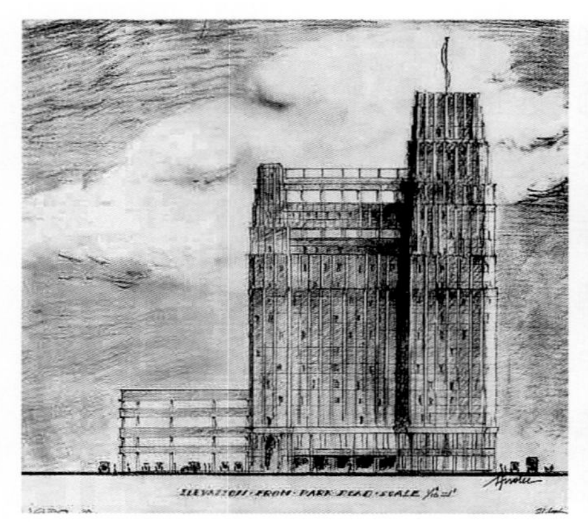

国际饭店见证了中国近现代建筑业的奇迹

Park Hotel witnessed a miracle in China's modern architectural industry

筑立面强调垂直线条，层层收进直达顶端，表现出美国现代派艺术装饰风格的典型特征。高耸稳定的外部轮廓，尤其是 15 层以上阶梯状的塔楼，在四周早已高楼林立的今天仍显得雅致动人。

国际饭店综合了当时世界各国的先进技术，更见证了中国近现代建筑行业的奇迹。中华人民共和国成立后，特别是 1980 年后，国际饭店历经多次重新装修，现为锦江集团旗下的四星级酒店。2001 年加拿大学者带来的邬达克档案馆的部分珍贵历史图片，至今仍在大堂夹层的走廊内展出。国际饭店于 2006 年被公布为全国重点文物保护单位。同年入选“首批中国 20 世纪建筑遗产”名录。

Reportedly, many people let their hats drop to the ground for looking up at it. It corroborates the reputation of the Park Hotel as the "Tallest Building in the Far East". During more than half the century when it remained the highest building in Shanghai, the centre of its rooftop flagpole was referred to as the "Zero Coordinate Point of Shanghai" for the municipal survey.

Park Hotel, also known as the Shanghai Joint Savings Society, stands 83.8 metres and 24 floors tall with 22 floors above ground and another 2 underground. It remained the tallest building in Shanghai and in China for 53 years until the Shanghai Hotel was built in 1983. By 1930, the Shanghai Joint Savings Society, co-founded by Kincheng Bank, the China & South Sea Bank, the Continental Bank and Yien Yieh Commercial Bank, decided to invest in a modern high-rise hotel, as the land price skyrocketed and real estate became a lucrative business. The location was selected at the intersection of Park Road (now Huanghe Road) and Bubbling Well Road (now West Nanjing Road), hence its name, "Park Hotel".

The hotel is undoubtedly the best and the most influential building designed by Ladislav Hudec. He won the bid for the project because of his bold innovation in high-rise structure and technologies that helped resolve the issue of soft ground settlement in Shanghai. He used a raft foundation comprised of 400 stakes, each 33 metres long, and reinforced concrete and an alloy steel upper structure that is both light and strong. Thanks to such choices, the Park Hotel suffered the least settlement among all the high-rises built in that period in Shanghai.

On the first floor was once the business hall of the Shanghai Joint Savings Society, in the basement the bank vault, and at the corner the hotel lobby. Such a layout was completely different from what it is now. On the second floor was a restaurant whose south side was comprised of large protruding French windows that offered an unobstructed view of the racecourse. Vertical lines are highlighted on its facades and shrink layer upon layer until the top, a typical feature of American modern Art Deco architecture. With a lofty and steady outline, especially the stepped tower on the 15th floor and above, the hotel looks particularly elegant and impressive compare to the surrounding modern skyscrapers.

Park Hotel exemplifies the best technologies from all over the world in that era and witnessed a miracle in China's modern architectural industry. After the People's Republic of China was founded, especially since 1980, the hotel has gone through multiple renovations. Today, it is a four-star hotel under Jinjiang International. Some of the valuable historical photos brought by Canadian scholars from a Hudec archive are still on display in the corridor of its lobby. In 2006, the Park Hotel was ranked as one of the National Key Cultural Relics Protection Sites. In the same year, it was listed among the "First Chinese Architectural Heritages in the 20th Century".

## 原四大公司
Former Four Great Department Stores

黄浦区南京东路
East Nanjing Road, Huangpu District

1917 年 10 月 20 日，位 于 南 京 路 浙江路口的先施公司开业，这是一座七层高的新古典主义风格大楼，英文名“Sincere”取意为“货真价实”，自此开启了南京路商业的新时代。

此后 20 年间，永安百货、新新公司和大新公司相继开业，四大公司既激烈竞争，又交相辉映。

1918 年 8 月，四大公司的第二家永安百货开业。以“环球百货”为号召，提供高档货品。大楼由公和洋行设计，是一座比先施更华丽的巴洛克风格建筑。为了在竞争中胜出，1932 年又在边上盖起一座 19 层高的摩天楼，与老楼以天桥相连。如今，这座简洁的钢构巨厦仍屹立在旁。

1926 年，第三家开业的新新公司建起一座更现代的装饰艺术风格建筑，由匈牙利建筑师鸿达设计。大楼设计有现代的线条，窗户很多，照明都是非直射的，还有舒适的电梯和充足的展示柜台。

1936 年，大新公司亮相。与前三家商店相比，这家规模最大，设备最新。大楼呈现代派风格，局部则带有装饰艺术风格以及中国民族风格的装饰，比如回字纹和顶楼的一排中国挂落。

四大公司浓缩了上海近代建筑风格的变迁，它们不仅整体设计精美，内部设施也十分完善，既有琳琅满目的商品，又开设有饭店、剧院、娱乐公园和屋顶花园等，大新公司还安装了中国第一台自动扶梯，非常类似于现代的购物中心。这四家的规模远远大于英资百货公司，顾客以国人为主，生意日渐红火，成为南京路的标志。

1949 年后，四大公司相继成为国营商店。1989 年被公布为上海市文物保护单位。如今，先施大楼由锦江之星酒店和上海时装商店共用，新新百货是第一食品商店旗舰店，大新公司（第一百货商业中心老楼）和永安百货在南京路新一轮城市更新中修缮改造，以吸引新时代的消费者。改造后的一百商业中心项目由一百老楼、新楼和东楼组成，通过多条跃层飞梯和空中连廊互通连接。新老大楼之间还加盖透明顶棚，仿佛撑起了一把巨大的透明伞。大楼之间的六合路作为南京路步行街的后街，也变身为新的城市休闲空间。

原永安公司

Wing On Department Store

原大新公司
The Sun & Co.

原先施公司
Sincere Co.

原新新公司
Sun-Sun Co.

On October 20, 1917, Sincere Co., located at the crossing of Zhejiang Road and Nanjing Road, opened. The seven-storeyed building is of neoclassical style, and its English name "Sincere" means "genuine goods at a fair price." And that heralded a new era of Nanjing Road commerce.

In the following 20 years, Wing On Department Store, Sun-Sun Co. and The Sun & Co. opened one after another. "The four great department stores" in Shanghai not only competed fiercely, but also complemented each other.

In August 1918, Wing On Department Store, the second of the four great department stores, opened. With the slogan of "Global General Merchandise", it offered highend goods. The building was designed by Palmer & Turner Group. It was a baroque style building more splendid than Sincere. In order to stand out from the competition, Wing On built a 19-storeyed skyscraper next to its old building and connected the two buildings by a flyover. Today, this simple steel structure is still standing by.

In 1926, Sun-Sun Co. started business in an even more modern Art Deco style building designed by Hungarian architect C. H. Gonda. The building is designed with modern lines, lots of windows, non-direct lighting, comfortable elevators and plenty of display counters.

In 1936, The Sun & Co. was founded. Compared with the previous three stores, this one was the largest in scale and boasted the latest equipment. The building presented a modern style, while decorative arts and Chinese national style decoration could also be found in it, such as the rectangular spiral design and a row of Chinese-style hanging fascias on the top floor.

The buildings of the four great department stores epitomize the changes of architectural styles in Shanghai in modern times. They were not only exquisitely designed, but were also equipped with up-to-date facilities. Apart from selling a wide range of commodities, the four companies also operate hotels, theatres, entertainment parks and roof gardens. The Sun & Co. also in-stalled the first escalator in China, which was very similar to a modern shopping centre. The scale of these four stores was far larger than those of British department stores in Shanghai. The customers were mainly Chinese, and with business booming day by day, they became the landmarks of Nanjing Road.

After 1949, the four great department stores became state-owned stores one after another. In 1989, it was designated as a cultural site protected by the city of Shanghai. At present, the building of Sincere is shared by Jinjiang Inn and Shanghai Fash-ion Store. The former Sun-Sun has become the flagship store of the First Food Mall. The buildings of The Sun (the old building of the Shanghai No.1 Department Store) and Wing On have been renovated in the waves of historic preservation of Nanjing Road to attract consumers of the new era. The Shanghai No.1 Department Store consists of old building, new building and East building after reconstruction, which are connected with each other through multiple flight ladders and air corridors. There is a transparent ceiling between the old and new building, which seems to be a huge transparent umbrella. And the Liuhe road between the building as the back street of Nanjing Road, also transformed into a new city leisure space.

# 1933 老场坊

1933 Old Millfun

虹口区沙泾路 10 号，原上海工部局宰牲场
10 Shajing Road, Hongkou District,
Former Site of Shanghai Municipal Council Slaughterhouse

1933 老场坊前身为工部局宰牲场，由建筑师卡尔·惠勒、结构工程师温特贝格设计。一个“正常”的建筑，每一个细节都来自现实条件的约束，约束愈多、对空间的牵扯愈多，一个本来平常的物体便愈发变形。宰牲场有着诡谲的空间，宛如一台协同运转的大机器，具有完整的流线。

宰牲场建成于 1933 年，建筑面积有 3.17 万平方米，是当时远东最大的现代化屠宰场。建筑由两部分组成：处理病死牛的独立焚烧楼，占地约 2200 平方米；四层主体建筑，畜牲楼包含畜养、宰杀、检疫、废弃物处理、冷库、市场等综合功能，占地约 8000 平方米，其中冷库和市场因建设九龙宾馆而被拆除。宰牲场形态是外方内圆，四周的方形空间是牛、羊、鸡兔的畜养处。中间的圆形容纳了与电动宰杀相关的所有功能，方圆之间形成了露天中庭，上空悬挂着 13 座天桥，在今天看来平添了空间的丰富层次。中庭开敞，在如此充满血腥的场所尽可能增强采光通风很重要，动物也会减少因宰杀叫声造成的恐惧。另外屠宰人员也可以通过天桥到达中心交通核，底层还有专用的牛坡道，避免了各宰牲区之间的干扰。废弃物通过管道，“零能源”滑落到半地下毛皮处理室，利用重力作用协助分类、清理、运输，是非常聪明的妙笔。

宰牲场有高标准的耐腐、卫生防疫要求，主要选用了无梁楼盖，包括四边形和八边形两种伞形柱。结构高度降低，节省了造价。顶面无梁，趋于平整，对清洁卫生是有利的。它多年高强度地使用，1949 年后经历多次改造仍留下完整、结实的钢筋混凝土结构体系，可见昔日的设计要求和施工质量非常优异。

宰牲场功能多变。2005 年，其价值被再度发掘，列入上海市第四批优秀历史建筑名单。2007 年完成改建工程，定位是创意产业园，更名为 1933 老场坊，从此蜚声海内外。2019 年被公布为全国重点文物保护单位。

As the predecessor of the 1933 Old Millfun, the former Shanghai Municipal Council Slaughterhouse was designed by architect Carl Wheeler and structural engineer Winterberg. When a "normal" building is constrained by the realistic conditions in every detail of its design, it will become increasingly deformed with more space constraints. This explains the weird space layout of the slaughterhouse, which looks like a completely streamlined big machine that enables coordinated operation.

Built in 1933 and with a floor area of 31,700 square meters, the slaughterhouse was the largest modern slaughterhouse in the Far East at the time. It is divided into two parts. One is an independent incineration building to deal with sick and dead cattle, covering an area of about 2,200 square meters. The other is a fourstorey main building for livestock with functions such as livestock raising, slaughtering, quarantine, waste disposal, refrigeration storage and marketing, and covering an area of approximately 8,000 square meters. The refrigeration storage and market areas were demolished for the construction of Shanghai Greenland Jiulong Hotel. The slaughterhouse is square on the outside and round on the inside, with the sounding square space to raise animals such as cows, sheep, chickens, rabbits, etc. The round area in the middle offered all functions related to electric slaughter. There was also an open-air atrium with 13 over-passes above it, adding rich space layers. Besides, the open-air atrium could also increase lighting and ventilation as much as possible in such a gory place, which was very important, and reduce the fear caused by animals' screaming when they were slaughtered. What's more, the butchers could reach the central traffic area through the overpasses. There was a special slope for cattle on the ground floor to avoid interference between different slaughter areas. The waste fell through the pipeline to the semi-underground hair and fur treatment room with no energy consumption. With the help of gravity, the waste was sorted, cleaned and transported, which was really a smart design.

The slaughterhouse meets the high standards of anti-corrosion, sanitation and epidemic prevention. It mainly uses beamless floor coverings, including quadrilateral and octagonal umbrella columns, which lowers the structure's height and saves cost. The top surface without beams is almost flat, which is conducive to cleanliness and hygiene. It was intensively used for many years. Despite many renovations after 1949, the complete and strong reinforced concrete structure still remains, which speaks to its high design requirements and excellent construction quality.

The slaughterhouse has variable functions. In 2005, its value was rediscovered and it was included in the fourth List of Excellent Historical Buildings in Shanghai. In 2007, the reconstruction of the project was completed, positioning it as a creative industrial park and renamed 1933 Old Millfun. Since then, it has gained fame, both at home and abroad. In 2019, the former site of the slaughterhouse was announced as a Major Historical and Cultural Site Protected at the National Level.

## 上海造币厂
Shanghai Mint

普陀区光复西路17号，原中央造币厂
17 West Guangfu Road, Putuo District, Former Central Mint

上海造币厂的前身中央造币厂是近百年来上海乃至中国制币工业发展演变的重要见证，创建至今，一直在进行钱币生产。它的诞生与上海金融中心地位的形成息息相关。

中央造币厂旧址现存铸币厂房（办公楼）、财政部部库旧址、水塔、原厂门门柱等历史建筑。其中铸币厂房（办公楼）于1922 年建成，由姚新记营造厂承建，钢筋混凝土结构，整体呈三段式结构，中部三层两侧两层，入口门廊有爱奥尼亚式柱支撑，显现出巴洛克风格。财政部部库旧址于 1935 年建造，为建筑师庄俊设计监造，大元建筑公司承建，库房分为两大间，内装有库门三座，库门采用对字转锁，并配备时间锁轮机关。

1949 年后，中央造币厂更名为人民造币厂。同年，厂修理间、电器间、物资库与上海人民印刷一厂修理间合并，成立上海人民铁工厂。1950 年，上海人民铁工厂改为上海人民印刷厂铁工分厂，至 1954 年，中国人民银行命名铁工分厂为国营六一四厂，也就是老上海人熟知的“614 厂”。1992 年，六一四厂改名为上海造币厂，2008 年，改为上海造币有限公司。

2014 年，中央造币厂旧址被公布为上海市文物保护单位。2019 年，中央造币厂旧址入选“第二批中国工业遗产保护名录”；2020 年，中央造币厂旧址入选“第三批国家工业遗产名单”，并迎来 100 周年庆典。

As the predecessor of the Shanghai Mint, the Central Mint is an important witness to the development and evolution of the mint industry in Shanghai and China over the past 100 years. Since its inception, it has been producing coins and is closely related to Shanghai's position as a financial center.

The mint premises (office building), the former site of the Ministry of Finance's warehouse, the water tower, gate posts of the former factory, etc. still remain. Among them, the office building was built by Yaoxinji Construction Factory in 1922. The reinforced concrete structure has three sections with three floors in the middle and two floors on both sides. The entrance porch is propped by Ionian columns, displaying a Baroque style. Built in 1935, the former site of the Ministry of Finance's warehouse was designed and supervised by architect Zhuang Jun and built by Dayuan Construction Company. The warehouse is divided into two large parts, and three doors with combination locks are installed inside, with the time wheel locking mechanism.

After 1949, the Central Mint was renamed People's Mint. In the same year, its repair room, electrical equipment room, and material warehouse and the No.1 repair room of Shanghai People's Printing Factory were merged to form Shanghai People's Ironworks. In 1950, Shanghai People's Ironworks became Shanghai People's Printing Factory Ironworks Branch, which was transformed in 1954 by the People's Bank of China into the state-owned 614 Factory widely known to old Shanghainese. In 1992, the 614 Factory was renamed the Shanghai Mint, which became Shanghai Mint Co., Ltd. in 2008.

The former site of the Central Mint was designated as a cultural site protected by the city of Shanghai in 2014. It was included in the second List of Industrial Heritage Sites in China in 2019 and the third list in 2020. It celebrated its 100th anniversary in 2020.

# 杨树浦水厂

Shanghai Water Works

杨浦区杨树浦路 830 号
830 Yangshupu Road, Yangpu District

如果从 1881 年英商上海自来水公司的筹建算起，到 2020 年，杨树浦水厂已经走过近 140 个年头了。作为全国最早成立的现代化自来水厂，杨树浦水厂不仅是重要的工业遗产，也见证了上海乃至中国自来水事业的发展历程。

杨树浦水厂的创办要追溯到 1875 年，立德洋行洋商格罗姆、立德尔、华脱司和邱裕记筹资白银 3 万两，在今杨树浦水厂南部购地 115 亩，开设供水公司，建成小型自来水厂。1881 年，水厂出售给筹建中的英商上海自来水公司，公司投资 12 万英镑在原址上予以新建。

1883 年，水厂建成供水，上海成为中国最早使用自来水的城市。为保证中心城区连续供水，公司在公共租界中心地点江西路香港路口建造了容量为 682 立方米、高 31.5 米的水塔一座，并敷设横越苏州河的供水管道。

为满足不断上升的供水需求，水厂开始扩建，1928 年，基本上形成现今的建筑格局。1952 年，水厂正式改名为上海自来水公司杨树浦水厂，简称“杨树浦水厂”。1989 年，杨树浦水厂被列为上海市文物保护单位；2003 年，利用厂内原储藏室和防空洞约 700 平方米的建筑面积，改造为上海自来水展示馆，向公众开放；2013 年，杨树浦水厂被公布为全国重点文物保护单位。

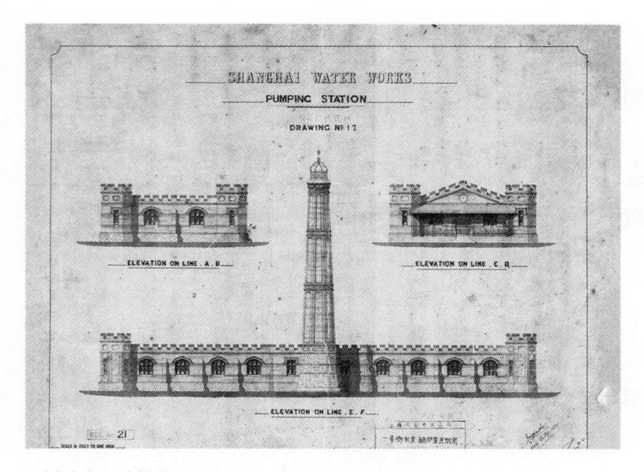


杨树浦水厂建筑的设计图
The design drawing of the Shanghai Water Works

杨树浦水厂见证了上海乃至中国自来水事业的发展历程
The Shanghai Water Works is a witness to the development of water supply in Shanghai and even in China

杨树浦水厂内的建筑基本为英国传统城堡形式，屋宇错落，建筑群布置和谐，体现出极强的形式美感。这些建筑设施多采用混凝土结构，是上海最早使用此类结构的建筑之一，在追求简约实用的工业建筑中极为珍贵，也是研究近现代工业建筑的重要案例。

杨树浦水厂是上海工业遗产最为重要的代表之一，时至今日，虽然在空间上经历了多次扩建，在制水技术上也经历了一次又一次的革新，水厂依然保留了原初的建筑格局，并且一直不间断地在进行制水工作，支持着上海供水事业的发展。可以说，杨树浦水厂是上海近代公共事业变迁的活化石。

By 2020, Shanghai Water Works will have existed for nearly 140 years if it is considered to have started in 1881, the year when the Shanghai Waterworks Co., Ltd. was founded in London. As the first modern waterworks built in China, Shanghai Water Works is not only important industrial heritage, but also a witness to the development of water supply in Shanghai, as well as China.

The establishment of Shanghai Water Works can be traced back to 1875, when F.A. Groom, A.I. Little, W.I. Waters and Church of Little & Co. raised 30,000 taels of silver and purchased 115 mu of land to the south of today's Shanghai Water Works to set up a water supply company and build a small water plant there. In 1881, the waterworks was sold to the Shanghai Waterworks Co., Ltd., which was under establishment and spent 120,000 pounds on building a new waterworks on the original site.

In 1883, when the construction was completed, Shanghai thus became the first city in China to use tap water. To ensure continuous water supply in the central urban area, the company built a water tower with a capacity of 682 cubic metres and a height of 31.5 metres at the junction of Jiangxi Road and Hong Kong Road at the centre of the Shanghai International Settlement, and laid a water supply pipe across the Suzhou Creek.

To meet the rising demand for water supply, the waterworks began to expand. In 1928, the waterworks was expanded in a style that it retains till this day. In 1952, the waterworks was formally renamed the Shanghai Waterworks Co., Ltd. Yangshupu Waterworks, or the Yangshupu Water Works for short. In 1989, Shanghai Water Works was designated as a cultural site protected by the city of Shanghai. In 2003, the store-room and air-raid shelter in the factory were transformed into the Shanghai Waterworks Museum, which was opened to the public. In 2013, the Water Works was announced as a

Major Historical and Cultural Site protected at the national level.

The buildings in Shanghai Water Works were designed to resemble English castles, with harmoniously scattered buildings reflecting the beauty of form. The buildings were among the first to adopt a concrete structure in Shanghai. They are of great value among the traditionally simple and practical industrial buildings, and also an important case study of modern industrial buildings.

An integral part of Shanghai's industrial heritage, Shanghai Water Works looks basically the same today despite all the expansions and changes in water supply technology. It has been contributing to the development of water supply in Shanghai continuously. Shanghai Water Works can be seen as a living fossil of the changing public utilities in modern Shanghai.

# 梧桐深处

Amidst the Plane Trees

# 中共一大会址
Site of the First National Congress of the CPC

黄浦区兴业路 76 号
76 Xingye Road, Huangpu District

1921 年 7 月 23 日晚，13 个操着各地方言的知识分子和两个外国人聚到一张长桌边，在一个不大的客堂间开启持续数日的讨论。这就是中国共产党第一次全国代表大会的开幕，大会宣告中国马克思主义政党的正式成立。见证会议的地方，就在那时的上海望志路 106 号（今兴业路 76 号）。

当年，中共一大会址被称为“李公馆”，是贝勒路树德里（今黄陂南路 374 号）的一部分。树德里为典型的石库门住宅，内有前后两排砖木结构的楼房，沿马路一排五幢石库门房屋，南北向分布，每幢一楼一底，独门出入，单元平面呈长条形，具有后期石库门里弄住宅的特点。这种民居，在一定面积的土地上，可以多安排几个单元，是适应当时城市土地紧缺、人口增长的需求。

此处外墙以清水青砖为主，镶以红砖水平带。山墙与硬山屋顶配合，为中西合璧的观音兜裙间跌落式。石库门门框为金山石条，黑漆大门，西式半圆形门楣，饰有盾牌花饰，带有巴洛克风格细部特征，体现了建筑中西合璧的风格特色。建筑门窗为传统的槅扇形式，窗下有精致的万字格木栏杆，内嵌贴金石榴花饰和贴金蝙蝠花饰。

1920 年秋落成后，李书城、李汉俊租用其中的望志路 106—108 号（今兴业路 76—78 号），并将这两幢一楼一底的沿街房屋后天井打通。

1921 年的夏天，各地中共早期组织代表以及共产国际代表来到“李公馆”的这个客堂间开会，深入探讨中国向何处去的命运。

20 世纪 50 年代初，中共一大会址经修缮成为纪念馆，至 1958 年又全面恢复原貌。1961 年 3 月，中共一大会址被公布为全国重点文物保护单位。

On July 23, 1921, thirteen intellectuals speaking different dialects and two foreigners quietly gathered at a long table in a humble living room and started a discussion lasting for a few days. This discussion in the name of the First National Congress of the CPC held at 106 Rue Wanz (now 76 Xingye Road) witnessed the founding of China's first Marxist political party.

In those years, known as "Li Residence", the building where the First National Congress of the CPC was held was in the Shude Lane, Rue Amiral Bayle (now 374 South Huangpi Road). Typical Shikumen community, Shude Lane had two rows of multi-story brick-and-timber townhouses. Along the streets running south and north, five buildings abutted each other. Every building had its own entrance with high brick walls enclosing a narrow front yard and a rectangular plane, which was a common characteristic of late Shikumen architecture. This kind of residence enabled more living spaces on limited plots, and was efficient to address the house shortage brought by the quick population increase in the city.

The fair-faced facade was mainly made of dark grey bricks, belted with red bricks, and inlaid with white plaster lines in the mortar joints. The door was painted black, with a pair of bronze knockers on it; the door frame was built with beige stone strips, and the arc-shaped part over the lintel was decorated with delicate Baroque flower patterns. The buildings had traditional latticework doors and windows, railings with Wan patterns and gilded pomegranate blossom and flying bat carvings were installed under the windows.

After the completion of Shude Lane in autumn 1920, Li Shucheng and his brother Li Hanjun rented 106—108 Rue Wantz (now 76—78 Xingye Road), and connected the rear patios of the two buildings along the street.

In the summer of 1921, representatives of early Communist

Organizations all over China and Comintern representatives came into the living room of this "Li Residence" to chew over the country's future.

In the early 1950s, after the site's restoration, a memorial hall was founded to commemorate the First National Congress of the CPC. It became Major Historical and Cultural Site Protected at the National Level in March 1961.

渔阳里
Yuyang Lane

黄浦区南昌路 100 弄 2 号
2 Lane 100, Nanchang Road, Huangpu District

老渔阳里建造于 1912 年，有老式石库门建筑八幢。这种里弄民居是在江南和上海县城民居传统建筑艺术的基础上，在外观上吸收部分西洋建筑艺术的装饰方法而形成的独特江南城市民居建筑。其中门牌为 2 号的房屋，曾是原安徽都督柏文蔚住宅，被称为“柏公馆”。它坐北朝南，屋面采用传统的小青瓦屋面，原始外墙为青砖墙，石库门头门楣为红砖砌筑，青石雀替装饰，门楣中包含精美山花；金山石材门框，正门为乌漆厚木门，配铜门环。杉木窗外设有木百叶窗，窗上是红砖砌木梳背券。

1920 年春，陈独秀从北京来到上海，不久入住环龙路老渔阳里 2 号，他主编的《新青年》也随迁于此。

6 月，陈独秀、李汉俊等在此开会，决定成立中国共产党早期组织。8 月，经征询李大钊意见，定名为共产党，陈独秀担任书记，这是党的发起组织，通过多种方式积极推动各地共产党早期组织的组建工作。而霞飞路新渔阳里 6 号（今淮海中路 567 弄 6 号），亦成为其开展工作的重要基地，后又成为中国社会主义青年团中央机关所在地。

1921 年 6 月，党的上海发起组筹备中国共产党第一次全国代表大会，老渔阳里 2 号为联络机关之一。中共一大后，陈独秀由粤返沪，仍居住于此。1922 年 10 月，中共中央局机关迁往别处。

1959 年 5 月，老渔阳里 2 号和新渔阳里 6 号均被公布为上海市文物保护单位。

今天，老渔阳里 2 号正式命名为“中国共产党发起组成立地（《新青年》编辑部）旧址”，新渔阳里 6 号现为中国社会主义青年团中央机关旧址纪念馆。

Old Yuyang Lane, built in 1912, has eight old-fashioned Shikumen buildings. This kind of Li-nong residence houses is unique in the city of Shanghai which combines the traditional Jiangnan and Shanghai township architectural style and some decoration methods employed in western architecture. The house with the door plate of No. 2, known as "Bai Residence of Bai", used to be the residence of the former governor of Anhui Province, Bai Wenwei. The south-facing twostorey, brick-wood building is roofed with traditional dark grey little tiles; the red-brick lintel of the door is decorated with blue-stone sparrow braces and delicate flower patterns. The door frame are made of granite; the wood door was painted black, with a pair of bronze knockers. Wooden sunblind was installed out of the Chinese-fir windows, which have comb-back arch made of red bricks upon them.

In the spring of 1920, Chen Duxiu came to Shanghai from Beijing, and soon moved to No. 2, Old Yuyang Lane, Route Vallon (now No. 2, Lane 100, Nanchang Road), and *La Jeunesse*, which he edited, also moved here.

In June, Chen Duxiu, Li Hanjun and others held a meeting here and decided to establish the early Communist organization in China. In August, after consulting Li Dazhao, the organization was named the "Communist Party". Chen Duxiu served as the secretary. This was the Party's initial organization which actively promoted the early Communist organization building across China by various means. No. 6 New Yuyang Lane, Avenue Joffre (now No. 6, Lane 567, Middle Huaihai Road) became an important base for its work, and later became the location of the Central Committee of the Chinese Socialist Youth League.

In June 1921, the Party's initiating team in Shanghai took No. 2 Old Yuyang Lane as the liaison office to prepare for the First National Congress of the Communist Party of China. After the founding of the

Party, Chen Duxiu returned from Guangdong and continued to live here. In mid-October 1922, the Central Bureau of the CPC moved elsewhere.

In May 1959, both No. 2, Old Yuyang Lane and No. 6, New Yuyang Lane were announced as cultural relics protected in Shanghai.

Today, No. 2, Old Yuyang Lane is officially named "The Preparation Office of the First National Congress of the CPC (the Former Editorial Office of *La Jeunesse*)", and No. 6, New Yuyang Lane the memorial site of the Central Committee of the Socialist Youth League of China.

## 步高里

Cité Bourgogne

步高里是上海较有名气的一个石库门里弄，它的弄道呈典型的鱼骨形：纵向主弄直贯南北，如同椎骨；北端与横主弄交会处是一个小广场，如同鱼头；鱼身横贯以三道支弄；支弄与主弄间以一道砖券划出空间界线，整体结构显得清晰有序。

步高里以中国牌楼式弄门为一大建筑特色，分别设于陕西南路和建国西路上的主弄入口，前者高大气派，上为装饰性斗拱与歇山顶式飞檐，下有一大两小三个简约的券门；后者小巧雅致，檐下以叠涩承托，门洞方正，与弄同宽。两座牌楼上均有“1930”字样，并以中法双语书写弄名，中西合璧。

步高里弄内的石库门式样都是简约的清水砖饰设计，门框两侧壁柱顶部凹凸有致的抽象柱头，半圆形山花内含有现代装饰意味的席纹，种种细节，颇为耐看。

1989 年，步高里被列入上海市首批优秀近代建筑名单，成为市级文物保护单位。 2007 年，步高里成为文化广场、人文艺术社区的经典保护建筑景观之一。

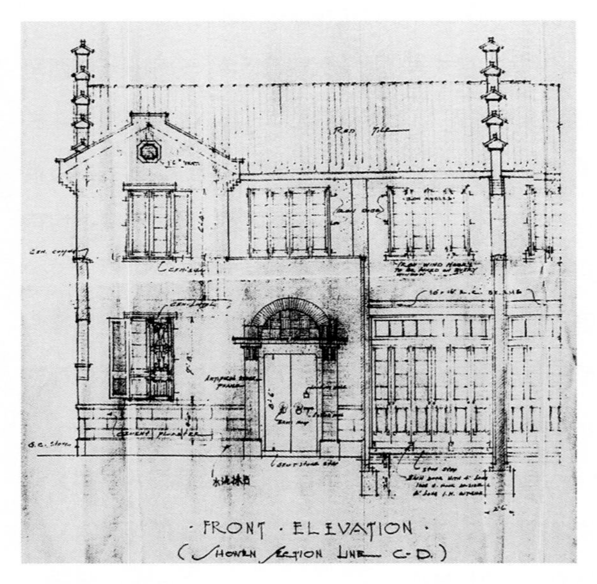


步高里石库门立面图原始稿

The original elevation design drawing of Bu Gao Li Shikumen

Cité Bourgogne, or Bu Gao Li, is a rather famous Shikumen neighborhood in Shanghai. The neighborhood is in the shape of fish bones: the main vertical passageway runs from south to north like the spine; at its north end where it crosses the main horizontal passageway, there is a small square like the fish head; three smaller passageways run horizontally, forming the fish body; and between them and the main horizontal passageway, there is a brick arch that works as the dividing line of a well-organized structure.

Cité Bourgogne is characterized by its two gates in the style of Chinese pailou. One is at the main entrance in the intersection between Shaanxi South Road and Jianguo West Road, looking lofty and magnificent with a decorative bracket, an East Asian hip-and-gable roof, with one large and two small minimalist arched doors; and the other is small and exquisite with supporting brackets under the eaves and square doors as wide as the passageway. Both gates are inscribed with "1930", and namesin Chinese and French, which are marks of western influence.

Cité Bourgogne is covered with minimalist fine bricks, the pilasters on both sides are topped with jagged and abstract chapiters; and the semicircular pediment features modernist cloth-like basket patterns inside.

In 1989, Cité Bourgogne, were listed as the first batch of outstanding historical buildings. Since then, it has become officially protected sites of Shanghai. In 2007, Cité Bourgogne became one of the protected classical buildings and sights of the Culture Square and of the culture and arts communities.

# 涌泉坊
Bubbling Well Lane

静安寺前的涌泉亭，曾为“静安八景”之一，亭中有井，昼夜沸腾，状似温泉，俗称“海眼”，1949 年后被加盖填塞。至今该寺附近的涌泉坊或许是与之相关的唯一纪念。涌泉坊位于愚园路 395 弄，建于 20 世纪 30 年代，为三层西班牙式的新式里弄典范。1989 年被公布为上海市文物保护单位。

涌泉坊占地面积 5300 平方米，建筑面积 6233 平方米。弄口是典型的过街楼，红色砖面外墙，大小拱门相连，大拱门正上方有一块白色横幅，从左往右书写“涌泉坊”三个红字，小拱门下方有雕刻精巧的花纹立柱。拱门上建有楼房，增加了利用空间，外立面装饰华美，显露出整片住宅区的高档品位。

坊内弄堂不多，仿佛带点高门大户的意味，建筑风格也略显错落。进来之后，六排建筑分列东西，主弄宽约 6.5 米，支弄约 3、4 米，米色的混凝土外墙，三层结构，围墙不到两米，阁楼和天台在三层之上。每幢房屋进深 15 米，前部为起居室、会客室，楼上为卧室，后部楼下为厨房、车库，楼上为保姆房间、贮藏室等。主要房间铺设檀木芦席纹地板，装修考究。

弄堂尽头 24 号洋房是华成烟草公司总经理陈楚湘住宅，人们称之为“陈家花园”。而陈楚湘便是涌泉坊的主要出资建造者。

小小洋楼上下四层，西班牙和意大利混合式建筑风格。屋顶高低错落，四个立面各具形制，有凸出的阳台，圆柱挑楼，外墙

涌泉坊建于20世纪30年代上海里弄住宅的黄金时期

Bubbling Well Lane was built in the 1930s, a golden age of the development of Shanghai Li-nong residence houses

贴彩色大面砖。整幢住宅共有 40 余间房，主楼铺设柚木地板。一楼为客厅、起居室、餐厅、厨房和车库。客厅畅亮，带有华美彩绘天花，其东面一间全堂紫檀木家具，称紫檀厅，西面一间为红木摆设，唤红木厅。二楼为卧室、女客专用室、客厅。三楼为子女卧室。一楼至三楼有宽敞的回旋式楼梯相连，大理石踏步，紫铜雕花扶手。四层一角有诵经堂。此建筑 1949 年后仍为居民住宅，至今是愚园路上一道独特的风景。

Yongquan Pavilion, which stood in front of the Jing'an Temple, was one of the "Eight Scenes of Jing'an". Inside the pavilion, there was a well, in which water bubbled day and night like a hot spring, hence the nickname the "Sea Eye". After 1949, the well was filled in. Today, Bubbling Well Lane, which is near the Jing'an Temple, is perhaps the only thing associated with the pavilion. Located in Lane 395, Yuyuan Road, Bubbling Well Lane was built in the 1930s, and it is seen as a model of the new three-storey Spanish-style lanes. In 1989, it was designated as a cultural site protected by the city of Shanghai.

Bubbling Well Lane covers an area of 5,300 square meters, with a floor space of 6,233 square meters. The entrance of the lane is a typical street building, with a red brick outer wall. A big arch is flanked by two smaller ones. Above the big arch is a white rectangle, on which the three characters of "Yong Quan Fang" are written from left to right in red. The smaller arches are underpinned by columns featuring delicately carved patterns. Rooms are built above the arches so that more space can be utilized. The exquisitely decorated façade reveals how fancy this residential area is.

There are not many alleys inside. The randomly distributed buildings seem quite independent. After you enter the lane, you will see six rows of buildings standing on the east and west sides. The main alley is about 6.5 meters wide and the side ones are about 3 or 4 meters. The three-storey structures have a beige concrete outer wall and the enclosure wall is less than two meters high. The attic and the rooftop are above the third floor. Each house is 15 meters deep. In the front, there is the living room and the reception room, and the bedroom upstairs; in the rear, there is the kitchen and the garage downstairs, and upstairs there is the nanny room, a storage room, and so on. In the beautifully decorated main rooms, the floor is made of reed-mat-patterned sandalwood.

At the end of the lane stands No. 24, a garden house was known as the Chens' Garden, it was once the residence of Chen Chuxiang, general manager of Huacheng Tobacco Company. Mr. Chen was the main sponsor of Bubbling Well Lane.

The four-storey building is a mixture of Spanish and Italian style. The roofs are high and low, and the four facades are of different shapes, with projecting balconies, a projecting floor supported by columns, and large colored tiles on the outer walls. The mansion has more than 40 rooms, and the main building has teak floors. On the first floor is the sitting room, living room, dining room, kitchen and garage. The spacious sitting room is decorated with a beautiful ceiling of colorful patterns. The east half of the room is called the Sandalwood Hall because of the red sandalwood furniture, and the west half is called the Rosewood Hall because of the rosewood decorations. On the second floor, there is a bedroom, a room for female guests, and a living room. Children's bedrooms are located on the third floor. The first to the third floors are connected by a spacious spiral staircase with marble steps and carved bronze handrails. There is a chanting hall in one corner of the fourth floor. The house remains a residential building after 1949 and it is an attraction on Yuyuan Road today.

新康花园
Former Jubilee Court

新康花园坐落在优雅的淮海中路南侧，复兴中路北侧，近汾阳路、宝庆路，与上方花园相邻。1989 年被公布为上海市文物保护单位。

新康花园北部 11 幢砖木结构花园住宅沿着弄堂两边分列，平缓的屋顶和花园墙的顶部都覆着红色的筒瓦，沿着屋顶的红色瓦片下檐边有连续的邮票齿形拱券纹。每幢两层住宅，一层一套，全部分层分户，上下住户各有独立门户进出，设计非常人性化。底层有外廊，套内房间为前后两排横向展开，前排中间为起居室，两边为卧室，均设有内阳台，后排为餐厅、厨房以及保姆卧室。外面还有汽车间，两层共享。外廊前方有一个近 40 平方米的花园庭院，庭院低矮的围墙上开设长方形镂空窗花，围有枪篱笆围墙。庭院内植雪松，苍翠挺拔，氛围幽静。二楼由西北角边门直上扶梯，起居室前为凹廊式大阳台，有红色筒瓦、铸铁阳台栏杆等典型构件，两根爱奥尼亚式螺旋形立柱和三段拱形凹入墙面，将整个阳台三等分。东西为卧室，后排是餐厅、厨房和保姆间，另有晒台。每层均有两个卫生间，每套建筑面积都在 200 平方米左右。这些建筑元素非常简练，彰显了西班牙式的特征。

新康花园，可谓当年最受青睐的理想居所；而大面积、多居室、带花园和汽车间等，都决定了其高级里弄公寓的定位。

Former Jubilee Court is in the south of elegant Middle Huaihai Road, north of Fuxing Road C., near Fenyang Road and Baoqing Road, and adjacent to the Shangfang Huayuan residential compound. In 1989, it was designated as a cultural site protected by the city of Shanghai.

The 11 masonry-structured garden villas in the north of the court lie on the two sides of the neighborhood. Their flat roofs and the tops of their garden walls are covered with red semicircle-shaped tiles underlined with running jagged arches. Every villa has two floors and each floor hosts one family with a separate entrance. In the front row is the living room. On either side of the living room, there is a bedroom with an internal balcony. In the back row is the dining room, along with the kitchen and the housekeeper's bedroom. There is a garage outside shared by both households. In front of the veranda, there is a garden courtyard of nearly 40 square meters. On its lower walls, there are rectangular, hollowed-out windows; whereas further outside, there are fences. The courtyard is planted with dark green and tall cedars, creating a tranquil environment. The second floor can be reached via a staircase by the side door on its north-western corner. In front of its living room, there is a large balcony in the form of a veranda. It features typical components of such a style, including red semi-circular tiles and iron railings, and is trisected by two Ionian columns and three recessed walls. The bedrooms lie in its western and eastern ends. In the back row is the dining room. There is also a flat roof. Each floor has two bathrooms and each suite has a floor area of around 200 square meters. Such minimalist architectural elements are very typical of Spanish buildings.

Former Jubilee Court was an ideal residential choice. With large villas comprising multiple rooms, gardens and garages, it is regarded as a senior condo complex.

# 武康大楼
Normandie Apartments

因为地处淮海中路、武康路、兴国路、天平路、余庆路的交叉口，契合基地的三角形形体独特巍峨，武康大楼一直被视为上海的城市文化符号之一。

武康大楼 1923 年由法商万国储蓄会开发，初名万国储蓄会霞飞路公寓，也称为诺曼底公寓或东美特公寓，设计者是还在美商克利洋行担任主创建筑师的邬达克。

这座八层高的大楼是沪上最早的外廊式公寓。平面因地制宜布置成熨斗形状，北面沿福开森路（今武康路）设置两个开口，以改善采光通风。底层为商铺，沿霞飞路（今淮海中路）设置拱廊式骑楼。居室大多朝南，户型结构灵活，有一至四室户之分，共有 63 套公寓和 30 多间用人房。内设电梯三部，消防楼梯多处。主入口门厅的两部电梯采用指针显示楼层。

大楼外观为法国文艺复兴风格，立面横向分三段：一二层基座为斩假石仿石墙面，中段三至七层用清水红砖，顶层檐部材质与基座相同，贯通的阳台和女儿墙构成双重水平线脚的檐部。底层半圆形拱券上设券心石，形成腰线的阳台采用宝瓶式石栏杆，三层和七层窗楣有山花装饰。

1943 年，福开森路更名为武康路，1953 年，诺曼底公寓更名为武康大楼。1995 年，武康大楼被列入上海市优秀近代建筑保护名录。2019 年，武康大楼通过架空线入地、合杆整治，露出了美丽的“素颜”。

Normandie Apartments, or Wukang Building, has been seen as a cultural symbol of Shanghai due to its lofty triangular foundation-fitting presence at the intersection of Middle Huaihai Road, Wukang Road, Xingguo Road, Tianping Road and Yuqing Road.

Developed by the International Savings Society (ISS), a French financial institution, in 1923, it was originally named I.S.S Avenue Joffre Apartments, or Normandie Apartments or Dong Meite Apartments. It was another signature work of Ladislav Hudec, who was also working with the American architectural office R. A. Curry as the chief architecture.

The eight-storeyed building is the oldest veranda-style apartment building in Shanghai. The building is appropriately designed into an iron shape to fit the plot. There are two entrances on the northern façade along Route Ferguson (now Wukang Road) to let in more light and air into the rooms. The ground floor is for shops, with arcades along Avenue Joffre (now Middle Huaihai Road). There are various types of apartments, from one room to four rooms, and most of the rooms face south. There are 63 apartments and more than 30 servants' quarters. The building has three elevators and multiple fire escape stairs. The two elevators in the entrance hall show the floor number with pointers.

The exterior of the building is in the French Renaissance style, and the facade is divided horizontally into three sections: the external walls on the first and second floors in the lower section are coated with artificial stones, while the third to the seventh floor in the middle section feature red bricks, and the eaves at the top are made of the same material of the lower section. The connected balconies and the parapets from eaves with double parallel corners. The semicircular arches on the first floor are equipped with keystones, and the balconies, which form a string course, are decorated with vase-shaped stone balusters. The window lintels on the third and seventh floors are decorated with pediments.

In 1943, Route Ferguson was renamed Wukang Road. The former Normandie Apartments was renamed in 1953 to Wukang Building. In 1995, the Wukang Building was included in the list of Outstanding Historical Buildings in Shanghai. In 2019, Shanghai carried out a city wide campaign to replace overhead cables with underground cables. The Normandie Apartments or Wukang Building restored its natural appearance, with the densely-woven cables that once long dominated the streetscape around the building removed and placed underground.

## 锦江饭店
Jin Jiang Hotel

黄浦区茂名南路59号
59 South Maoming Road, Huangpu District

如果你打听华懋公寓，至少有一半的上海市民，尤其是年轻人一时找不到方向。然而，1951 年起它的新名字“锦江饭店”在沪上却家喻户晓，更由于 1972 年《中美联合公报》的签署而蜚声海内外。

华懋公寓是 1925 年由华懋地产公司投资建造的传统哥特式建筑，钢筋混凝土框架结构，俗称“十三层”，高 57 米，建筑面积 21202 平方米。公寓外部以棕色面砖贴饰，方格钢窗排列齐整，石料窗框和垂直线条，强化了“拔地而起”的效果。公寓内设多部客货运电梯，出入便捷。11 层仿英国宫廷式大餐厅面积 800 平方米，皇家古典式吊灯金碧辉煌。典雅烛形壁灯、持盾武士浮雕，刚柔相济。

1929 年，华懋公寓与外滩高 77 米的华懋饭店（又称“沙逊大厦”，现为和平饭店北楼）同年竣工，成为当时沪西第一高楼。

华懋公寓地段好、品位高，两侧建有商店、影院等，开综合性建筑之端，销售顺畅。于是，五年后新沙逊洋行又在南边不远处建起一座标高 78 米、18 层的峻岭公寓。

1948 年，华懋公寓被“传奇女子”董竹君买下。1950 年，她将自己创立经营的锦江川菜馆、锦江茶室合并为锦江饭店。翌年 6 月，锦江饭店在华懋公寓开业，从此“华懋”改姓更名。

1959 年，锦江饭店建成小礼堂。随着几年后南楼的兴建，华懋公寓渐渐被称作锦江饭店北楼，中楼则是接待贵宾的峻岭公寓。

锦江饭店曾长期名列上海涉外宾馆规模之最。1972 年 2 月 27 日，《中美联合公报》在锦江饭店小礼堂内签署，这里见证了中美关系和中国走向世界的一个时代的开启。

华懋公寓、峻岭公寓与茂名公寓属上海市首批优秀近代建筑、上海市文物保护单位。

If you inquire about Cathay Mansions, at least half of Shanghai residents, especially young people, will not be able to give directions right away. However, since 1951, its new name, "Jin Jiang Hotel", has become a household name in Shanghai, and it has been known far and wide since it witnessed the signing of the *Sino-US Joint Communique* in 1972.

The Cathay Mansions is a traditional Gothic building invested in and built by the Cathay Real Estate Company in 1925. It is a reinforced concrete structure, commonly known as "the 13 Storeys", standing 57 meters tall, with a floor area of 21,202 square meters. The exterior wall is veneered with brown tiles. The square steel sashes are neatly arranged. The stone window frames and vertical lines strengthen the effect of "rising above the ground". The mansions are equipped with a number of passenger and freight elevators for convenient access. On the 11th floor, the big dining room, which imitates the English court style, covers an area of 800 square meters, lit by the resplendent royal-classical chandeliers. Elegant candle-shaped wall lamps and the relief of warriors show a perfect combination of force and grace.

In 1929, the Cathay Mansions and the 77-meter high Cathay Hotel (also known as "Sassoon House", now the North Building of Fairmont Peace Hotel) on the Bund, were completed in the same year, becoming the tallest buildings in west Shanghai at that time.

The high-end Cathay Mansions enjoyed a good location, with shops and cinemas built on its both sides. It marked the start of integrated building development and it was very well-received in the market. As a result, five years later, Messrs Sassoon Sons & Co built the 78-meter high Grosvenor House, with the 18-storeys, in the south not far away.

In 1948, the Cathay Mansions was bought by the "legendary woman" Dong Zhujun. In 1950, she merged the Jin Jiang Sichuan Restaurant and Jin Jiang Teahouse, she had established and operated, into Jin Jiang Hotel. In June the following year, Jin Jiang Hotel opened to business at the Cathay Mansions, giving the latter a new name and identity. In 1959, Jin Jiang Hotel built a small auditorium. With the construction of the South Building a few years later, the Cathay Mansion gradually began to be called the North Building of Jin Jiang Hotel, and the Middle Building was the Grosvenor House for VIP guests.

Jin Jiang Hotel has long ranked as the top hotel in receiving foreign guests in Shanghai. On February 27, 1972, the *Sino-US Joint Communique* was signed in the small auditorium of Jin Jiang Hotel, which witnessed the opening of a new era in Sino-US relations and China going global.

The Cathay Mansions, the Grosvenor House and the Mao Ming Building are among the first batch of heritage architecture of modern Shanghai and the Heritage Site under the Protection of Shanghai Municipality.

# 上海中山故居

Former Residence of Dr. Sun Yat-sen in Shanghai

黄浦区香山路7号
7 Xiangshan Road, Huangpu District

走近香山路 7 号，首先映入眼帘的是坐落在故居门前的孙中山坐式铜像，铜像四周是生机盎然的玉兰、香樟等树木花卉。先生神情从容坚毅，举止庄重安详，炯炯目光中充满了深情和希冀。

上海孙中山故居是一幢两层欧洲乡村式小洋楼，建于 20 世纪初，由当年旅居加拿大的华侨集资买下赠送给孙中山。

1918 年 6 月，孙中山辞去海陆军大元帅职务后从广州回到上海，与夫人宋庆龄入住香山路 7 号。他虽然屡遭挫折但毫不气馁，1918 年至 1920 年间，在这里总结革命经验，继续探索革命新路，先后撰写了《孙文学说》《实业计划》等重要著作。

孙中山在这里会见了中国共产党人李大钊、林伯渠以及共产国际代表马林、越飞等，商谈国共合作，为“联俄、联共、扶助农工”三大政策的确立和第一次国共合作的实现奠定了基础。

1924 年，孙中山在这里召开记者招待会，重申“北上宣言”，进一步阐明了北上谋求和平统一中国的主张。

1925 年 3 月孙中山逝世后，宋庆龄继续居住至 1937 年。抗日战争全面爆发后，宋庆龄移居香港、重庆。1945 年底，宋庆龄回到上海，将此寓所移赠给国民政府，作为孙中山的永久纪念地。中华人民共和国成立后，故居于 1961 年被列为首批全国重点文物保护单位，1988 年，正式对外开放。

故居坐北朝南，占地1013平方米，建筑面积452平方米。外墙装饰有深灰色鹅卵石，屋顶以洋红色鸡心瓦覆盖。现故居内的陈设绝大多数是原物原件，并根据宋庆龄生前的回忆，按20世纪二三十年代的原样布置。

紧邻孙中山故居还设有2006年11月正式开馆的孙中山文物馆，由一幢欧式洋房改建而成。文物馆共有三层，八个展区，陈列文物、手迹、史料300余件，向人们展示了孙中山为中华民族的伟大复兴所立下的不朽功绩。

Approaching 7 Xiangshan Road, the first thing that comes into sight is the brass seated statue of Sun Yat-sen in front of the former residence, surrounded by vibrant magnolias, camphor trees as well as other flowers and plants. The statue of Dr. Sun Yat-sen has a resolute look and a serene demeanour, with the two bright eyes full of love and hope.

The Yat-sen was bought by the overseas Chinese living in Canada and given to Dr. Sun Yat-sen as a gift.

In June 1918, Dr. Sun Yat-sen returned to Shanghai from Guangzhou after resigning as Generalissimo of the military Nationalist Government, and moved into 7 Xiangshan Road with his wife Soong Chingling. He was not discouraged in spite of repeated setbacks. It was in this building that he successively completed *Dr. Sun Yat-sen's Doctrine*, *The International Development of China* and other important works.

Sun Yat-sen met with not only such CPC members as Li Dazhao and Lin Boqu, but also the representatives of Comintern Maring (Henk Sneevliet) and Adolf Abramovich Joffe, which laid foundation for the establishment of the three policies of "aligning with the Soviets, cooperating with the Communists and helping the farmers" as well as the first Kuomintang-Communist cooperation.

In 1924, Sun Yat-sen held a press conference in this building, reiterating the "Statement on Travelling North," and further clarified the idea of going north for peaceful reunification of China.

After Sun Yat-sen passed away in March 1925, Soong Chingling continued to reside in the building until 1937. After the outbreak of the War of Resistance against Japanese Aggression, Soong Chingling moved to Hong Kong and then Chongqing. At the end of 1945, she returned to Shanghai and donated the residence to the Nationalist Government as a permanent memorial site of Sun Yat-sen. After the founding of the People's Republic of China, the former residence was listed as one of the first batch of Major Historical and Cultural Sites Protected at the National Level in 1961, and was officially opened to the public in 1988.

Facing south, the building covers an area of 1,013 square meters, with a floor area of 452 square meters. The exterior walls are decorated with dark grey cobblestones, and the roof is covered with magenta heart-shaped tiles. Today, most of the furnishings in the former residence are original ones, and are arranged in the same way as they were in the 1920s and 1930s based on the memories of Soong Chingling.

Next to the former residence is Shanghai Museum of Sun Yat-sen, another European-style building which was officially opened in November 2006. The Museum has three floors and eight exhibition areas, displaying more than 300 pieces of relics, manuscripts and historical materials, showcasing the ever-lasting achievements made by Dr. Sun Yat-sen for the great rejuvenation of the Chinese nation.

# 上海宋庆龄故居

Former Residence of Soong Chingling in Shanghai

徐汇区淮海中路1843号
1843 Middle Huaihai Road, Xuhui District

宋庆龄在上海的最后一处住所坐落于徐汇区淮海中路 1843 号。

1949 年春，宋庆龄在这里迎接了上海的解放。1963 年开始，宋庆龄只在过年过节时回上海居住。1978 年底，宋庆龄回这里过春节，居住了两个月，这是她最后一次回到上海住所。

1981 年 5 月 29 日，宋庆龄在北京逝世，上海住所作为故居对外开放，供人瞻仰。1981 年 10 月，故居被列为上海市文物保护单位。2001 年 6 月，被公布为全国重点文物保护单位。

淮海中路 1843 号的宋庆龄故居，主体建筑为一幢乳白色船形的假三层西式楼房，建筑面积 700 平方米。此建筑始建于 20 世纪 20 年代初，最初为一位希腊船主的私人别墅。建筑底层为客厅、餐厅、书房，二楼为宋庆龄卧室，办公室和保姆李燕娥的卧室。楼前有宽广的草坪，楼后是花木茂盛的花园，四周遍植常青的香樟树，环境优美清静。

故居书房中收藏着孙中山先生演讲的珍贵录音唱片和他亲笔题字的遗著。二楼卧室内的一套藤木家具是宋庆龄结婚时的嫁妆，沙发和茶几是孙中山先生使用过的。室内摆放着孙中山先生 18 岁时的照片和他逝世前一年与宋庆龄的合影，两位革命家的风采在此定格。

Soong Chingling's last residence in Shanghai was located at 1843 Middle Huaihai Road, Xuhui District.

In the spring of 1949, Soong Chingling welcomed the liberation of Shanghai. After 1963, Soong only returned to Shanghai on festive occasions. At the end of 1978, she came back for the Spring Festival and lived here for two months. That was the last time she returned to her residence in Shanghai.

On May 29, 1981, Soong passed away in Beijing, and her abode in Shanghai was opened to the public. In October 1981, the former residence was listed as a Monument under the Protection of the Shanghai Municipality. In June 2001, it was listed as a Major Historical and Cultural Site Protected at the National Level.

The main building of the Former Residence of Soong Chingling at 1843 Middle Huaihai Road is a white boat-shaped false three-storey Western-style building with a floor area of 700 square metres. The building was first built in the early 1920s as a private villa for a Greek ship owner. The ground floor of the building once served as the living room, dining room and study. Soong's bedroom, office and the Nanny Li Yan'e's bedroom were on the second floor. There is a wide lawn in front of the building, and a garden with lush flowers and trees behind. The environment amid evergreen camphor trees is very beautiful and quiet.

The study of the Former Residence houses precious recordings of Dr. Sun Yat-sen's speeches and posthumous works with his own inscriptions. The set of rattan wood furniture in the bedroom on the second floor was Soong's dowry when she got married, and the sofa and end-table were once used by Dr. Sun. A photo of Dr. Sun in his 18 and a group photo of him and Soong taken a year before his death has been placed in the room. The demeanour of the two revolutionaries is thus frozen in frame.

# 周公馆

Former Residence of Zhou Enlai

黄浦区思南路73号，中国共产党驻沪办事处旧址
73 Sinan Road, Huangpu District,
Former Site of the CPC Delegation Office in Shanghai

走进法国梧桐遮天的思南路，那幢墙上覆满了爬山虎，带竹篱笆围墙的红顶小楼就是“周公馆”，它是解放战争时期中共代表团驻沪办事处旧址。

周公馆原为欧洲近代独立式花园住宅，占地面积 449 平方米，建筑面积 625 平方米，四层砖混结构，南立面中部前出为内阳台，一楼有外置直梯径往花园；北立面中部凸出，二楼设混凝土结构的阳台；底层为辅助用房，饰尖券门廊。建筑外墙为水泥砂浆抹层嵌天然砾石饰面，赭红漆木百叶窗，红陶机制平瓦双坡顶、山墙部分做跌檐处理；主入口西向，有外置石阶直抵一楼。入口处红砖砌筑的门楣及室内壁炉、楼梯望柱、彩色镶嵌玻璃等具有德国青年风格派的艺术特征。办事处设立后，住宅根据当年具体情况进行了改造。

周公馆所在的“义品村”则由 23 幢三层（局部四层）独立式花园住宅组成，位于思南路东侧 51—95 号（单）。该片区建筑统一中稍有变化，花园较大，绿化完整，配套设施齐全。

1959 年，思南路 73 号被公布为上海市文物保护单位。1993 年，“义品村”花园住宅被公布为上海市优秀历史建筑。2019 年，中国共产党代表团驻沪办事处旧址被公布为全国重点文物保护单位。如今，这一带以“思南公馆”之名为人所熟悉。

On Sinan Road in leafy shadows of plane trees, there is a small red-roof little house surrounded by bamboo fences and with its wall covered with Japanese Ivy known as “Residence of Zhou Enlai”.

Residence of Zhou Enlai was an unattached garden house of modern European style, which covers an area of 449 square meters with a floor area of 625 square meters. It is four-storey brick-concrete building, whose south façade has an inner balcony in the middle and an outer stair leading directly into the garden. The north façade has a bulk and a concrete balcony on the second floor. On the ground floor are auxiliary room and a pointed-arch porch. The exterior walls of the building are plastered with cement mortar layer inlaid with natural gravels. Wooden sunblind is painted Indian red; the double-pitch roof is covered with machine-made redware flat tiles; and the building has jerkinhead gables with the end of the roof hipped for only part of its height. The main entrance is facing the west, having an outer stone stair leading to the first floor. The red-brick lintel at the gate, the in-house fireplace, the balusters of the stairs, and the colorful mosaic glass are all of the artistic characteristic of the German Jugendstil style.

“Yipin Village” at No. 51-95 (odd numbers) on the east side of Sinan Road where the Residence of Zhou Enlai is located has 23 three-storey (partially four-storey) unattached garden houses. In this block, buildings are mostly identical with some slight variations.

In 1959, No. 73 Sinan Road was announced as a Monument under the Protection of the Shanghai Municipality. In 1993, “Yipin Village” garden houses were announced as Heritage Architecture in Shanghai. In 2019, it was announced as a Major Historical and Cultural Site Protected at the National Level. Today, this area is known as “Sinan Mansions”.

## 中福会少年宫

The China Welfare Institute Children's Palace

静安区延安西路64号
64 West Yan'an Road, Jing'an District

延安西路 64 号，是一座建在大草坪中的白色宫殿，那浑然一体的大理石饰面和舒展的希腊复兴式侧翼，尤其令人惊叹，这就是被人们称为“大理石宫殿”的嘉道理别墅。

别墅内外墙面及地面几乎都饰以大理石，入口处有爱奥尼式大理石柱廊，大厅顶部为大理石穹顶，就连楼梯的踏步、扶手和栏杆也饰以大理石，且均由意大利进口。

1919 年，原主人英籍犹太人艾利·嘉道理在黄陂南路的旧宅失火损毁后，便买下犹太总会建造的大西路 64 号作为私宅，并设计重建。1924 年，嘉道理一家入住这座宛如宫殿的豪宅。1929 年，在原两层上又加建一层，至 1931 年全部建成。

嘉道理别墅采用希腊古典复兴风格，多层次水平线条的处理，使建筑颇为舒展。平面呈矩形，中心是大舞厅，穹顶上饰精美浮雕，可容数百人翩翩起舞。东西两侧分别布置宴会厅、会客厅、餐厅、棋牌室等。二楼为卧室，房间共计 20 余间，装修各异。别墅南侧有近万平方米的大草坪，建有暖窑花房、马厩、鹿厩、网球场，由铁栅围墙围成一个私密的后花园。这座耗资 100 万两白银的别墅，超过了哈同花园，成为当时上海最贵的花园洋房之一。

二战前夕，大批犹太人涌入上海，嘉道理作为上海犹太总会会长，成立“上海援助欧洲难民委员会”，在别墅里为犹太同胞提供庇护。太平洋战争爆发后，别墅被日军侵占，用作军事机构，而嘉道理则一直被软禁在这里直至病逝。

1945 年日本投降，这里就成了美国军人的活动中心。1953 年，宋庆龄创办的少儿图书馆和少儿文化馆合并，一起迁入嘉道理别墅，取名中国福利会上海市少年宫。

一个最初按照公共建筑设计、后又反复被用作公共建筑的私家别墅，承载了近现代上海的沧桑巨变。曾经是一个犹太家族财富象征的私家宅院，今天变成了孩子们欢乐的天堂。2019 年，中福会少年宫被公布为全国重点文物保护单位。

Located at 64 West Yan'an Road, the former residence of Kadoorie is a white palace with a large lawn. The marble finishes and the unfolding Greek Renaissance-style wings are particularly impressive, hence came the name of "Marble Mansion".

Almost all the interior and exterior walls and floors of the building are decorated with marble. There are marble colonnades of the Ionic order at the entrance. The top of the hall features a marble dome. Even the steps, handrails and railings of stairs are decorated with marble, all imported from Italy.

In 1919, after his old house at South Huangpi Road was damaged by fire, Elly Kadoorie bought 64 Great Western Road built by the Shanghai Jewish Club as his private house. Later, he redesigned and rebuilt the house. In 1924, the Kadoorie family moved into this palace-like mansion. In 1929, another storey was added on top of the original two storeys, and the construction was completed in 1931.

The Former Residence of Kadoorie adopts the Greek Revival style, and the multilevel horizontal line treatment unfolded the building to a large extent. The house is a rectangular building, with a large ballroom in the centre, which can hold hundreds of people. The dome is decorated with exquisite embossment. Banquet hall, reception hall, dining room, chess and card room are arranged in both wings of the building. There are more than 20 bedrooms on the second floor with different furnishing styles. On the south side of the building lies a big lawn covering nearly 10,000 square metres, which is a private back garden surrounded by grill walls. There is a greenhouse with warm kilns, a horse stable, a deer pen and a tennis court in the garden. The majestic villa cost much more than Hardoon Garden (1,000,000 taels of silver), becoming one of the most expensive garden houses in Shanghai at that time.

On the eve of World War II, a large number of Jews swarmed

into Shanghai. As the president of Shanghai Jewish Club, Kadoorie established the "Committee for the Assistance of European Jewish Refugees in Shanghai (CAEJF)" to provide shelter for the Jews in his residence. After the outbreak of the Pacific War, the place was occupied by the Japanese army for military use, where Kadoorie was kept under house arrest until his death.

When Japan surrendered in 1945, it became the activity centre of American soldiers. In 1953, Children's Library and Children's Cultural Centre founded by Soong Chingling merged and moved into the Former Residence of Kadoorie with the name of CWI Children's Palace.

A private house, originally designed for public use and repeatedly used as a public building later, has witnessed the vicissitudes of Shanghai in modern times. The private house of a moneyed Jewish family in history has now become a paradise for children. In 2019, it was announced as a Major Historical and Cultural Site Protected at the National Level.

# 长宁区少年宫
Shanghai Changning Children's Palace

长宁区愚园路1136弄31号，原王伯群住宅
31 Lane 1136, Yuyuan Road, Changning District, Former Residence of Wang Boqun

长宁区少年宫内，有一幢精致漂亮的建筑，即王伯群住宅。

王伯群住宅占地十亩，南面临愚园路，大门开在愚园路1136弄，汽车可以通过这条路直接驶入宅邸。住宅由美商协隆洋行的中国建筑师柳士英设计，为以英国哥特式复兴风格为主的“折衷主义”建筑。住宅为四层砖混合结构。建筑主立面正面以中间为纵轴线，两侧形成对称；纵轴线两侧设计有对称的石扶梯，步上扶梯可进入二楼的正客厅。客厅的面积相当大，除了会客外，还可充当宴会厅、舞厅、酒吧等。三楼以上则是主人用房，从东门进入大厅的过道设计有主、次卧室，书房，会客厅等。底层则作为工作人员用房。建筑的背面（北面）及东、西两侧设计为仿欧洲中世纪城堡式样，以深褐色的面砖作为外墙贴面，显示出古城堡的苍劲、古朴，窗户多用斩毛假石镶边，窗顶设计为四心尖券，确实有欧洲中世纪城堡的印痕。

1949年后，该宅曾作为部队机关和长宁区政府机关所在地。1960年，区政府迁新址，该宅改作长宁区少年宫。1989年被公布为上海市文物保护单位。

The Former Residence of Wang Boqun covers an area of 10 mu. It faces Yuyuan Road in the south. The front gate is opened in Lane 1136 of Yuyuan Road, through which vehicles could drive directly into the residence. The house was designed by Liu Shiying, a Chinese architect of the American Company Fearon, Daniel & Co. and is a so-called eclectic building mainly of British Gothic Revival style. The building is a four-storeyed brick-concrete structure. The main façade of the building takes the middle as the longitudinal axis, forming symmetry on both sides; there are symmetrical stone stairs on both sides of the longitudinal axis, leading to the living room on the second floor. The living room is very spacious and can also be served as banquet hall, ballroom, or bar in addition to receiving guests. From the third floor and above are the master's rooms. The master room, bedrooms, study, and drawing room, etc. are on both sides of the corridor from the east door to the main hall. The ground floor is for staff use. The back (north) as well as the east and west sides of the building are designed to imitate the European medieval castle style. The dark brown facing bricks are used as the external wall veneer, which gives the building a bold and primitive touch of an ancient castle. The window frames are mostly inlaid with chopped stones. The top of the windows is designed as four-pane pointed arch, which is characteristic of medieval European castles.

After 1949, the property was successively used as the office building of the army and Changning District government. In 1960, the Changning District government moved to a new site, and this place has been used as Shanghai Changning Children's Palace since. In 1989, it was designated as a cultural site protected by the city of Shanghai.

# 上海工艺美术博物馆

Shanghai Museum of Arts and Crafts

徐汇区汾阳路79号
79 Fenyang Road, Xuhui District

沪西的汾阳路由法租界公董局筑于 1902 年，法文路名 Route Pichon，中文名毕勋路，以法国驻华公使毕勋命名。79 号内有一幢大楼，通体以白色石材作为装饰。

1912 年，旅沪法国侨民法诺、盘滕、麦地，以及中国人章鸿笙等人合作创办了万国储蓄会，向法国领事馆注册为公共性商事机构，它也为中国出现的第一家商业储蓄机构，也是旧中国最大的储蓄机构。万国储蓄会利用储蓄金投资上海的房地产，是旧上海最大的房地产商之一。大股东要分红，高层管理者要提薪，于是，万国储蓄会除了兴建商品住宅外，还建造了一批豪宅，以租金代分红，提供给储蓄会大股东和高级管理人员使用。

汾阳路 79 号住宅建成于 1925 年，美商克利洋行设计，当时的克利洋行雇员邬达克参与了设计和监造。住宅占地约 3900 平方米，主建筑占地面积约 420 平方米，假四层砖石混合结构，属法国后期文艺复兴建筑样式。以朝南一侧为主立面，面对大块的草坪，主立面平面呈“凸”字形，中间凸出的部分形成半圆，并作为中轴线，两侧退进呈平面，严格对称，主立面的底层与二层之间设计左右对称的环形石阶，二层设计阳台，二层凸出的部分是主客厅，与平台相接，建筑的连贯性很好；在底层的两扶梯之间设计有半球，在这里设计一喷泉，装饰效果极佳。北面是建筑的背面，设计有高大的雨棚，汽车可以直接驶入雨棚而进入大楼。建筑的二层设计为罗马拱券长窗，三层则为方窗，层与层之间设计有明显的腰线。

1949 年后，由上海市人民政府接管。1963 年，上海市工艺美术研究所迁入此楼。1989 年被公布为上海市文物保护单位。如今，这里是上海工艺美术博物馆所在地。

Fenyang Road in the west of Shanghai was built by Conseil d'Administration Municipale de la Concession Française de Changhai, Municipalité Française in 1902 with the French name Route Pichon after Stephen Jean-Marie Pichon, the French Minister to China. There is a building in the Residence at No. 79, fashioned with white stones.

In 1912, French settlers in Shanghai Rone Fano, Jean Beudin, and Henri Madier, as well as Zhang Hongsheng and other Chinese businessmen, co-founded the International Savings Society (ISS) and registered it at the French Consulate-General as a public commercial organization. It also became the first commercial savings agency in China and the largest one at the time. The agency used the savings to invest in real estate in Shanghai, and became one of the largest real estate developers in old Shanghai. To meet the needs of dividend sharing among major shareholders and salary promotion of senior managers, in addition to developing commercial residential buildings, the ISS also built a number of luxury houses rented to these major shareholders and senior management, and thus the dividends and salary promotion were offset by the corresponding rentals.

The Residence at 79 Fenyang Road was completed in 1925 with the design of the American company Curry R.A, and noted architect Laszlo Hudec was involved in the design and supervision as its employee. The Residence covers a land area of about 3,900 square metres, with the main building taking up an area of about 420 square metres. It is a false four-storey masonry structure in the late French Renaissance architectural style. The main façade is facing south in the shape of an upturned "T", with a large lawn in front of it. The protruding part in the middle forms a semicircle, and serves as the central axis. The two sides are recessed in plane and strictly symmetrical. Circular stone steps are designed in a symmetrical manner between the bottom layer of the main façade and the second floor, and the balcony is designed on the second floor. The protruding part of the second floor is the main living room, which is connected with the platform, forming great coherence; a vertically-retracted hemispheric structure is designed between the two stairs on the ground floor, where a fountain is furnished to show excellent ornamental effect. The north side is the back of the building, where a tall and big rainshed is designed. One can drive the car directly through the rainshed and enter into the building. The second floor of the building is designed with long windows in Romanesque arch, and the third floor is designed with square windows. Obvious waist lines can be seen between the storeys.

After 1949, the Residence at 79 Fenyang Road was taken over by the Shanghai Municipal People's Government. In 1963, the Shanghai Arts and Crafts Research Institute moved into the building. In 1989, it was designated as a cultural site protected by the city of Shanghai. Today, it is the site of the Shanghai Museum of Arts and Crafts.

# 上海市城市规划设计研究院

Shanghai Urban Planning & Design Research Institute

静安区铜仁路333号，原吴同文住宅
333 Tongren Road, Jing'an District,
Former Residence of D.V. Woo

2014 年 6 月 14 日是中国第九个文化遗产日，虽然上海上午的气温已经达到 30 多度，铜仁路 333 号周边还是蜿蜒排起了 300 多米的长队，开放仅三小时，就有超过 1200 名市民涌入改造后首次开放的“绿房子”先睹为快，正如建筑师当年向业主所夸耀的：“我为你设计的住宅将 100 年都不落后。”

吴同文住宅是著名建筑师邬达克设计成熟期的代表作，经典的现代风格花园洋房。建筑面积近 1700 平方米，因为外立面和围墙均采用绿色釉面砖，俗称“绿房子”。今为上海市优秀历史建筑。

吴同文是旧上海著名的颜料商，因靠军绿色颜料致富，故视绿色为幸运色。该地块是贝家小姐的陪嫁礼，毗邻的哈同路（今铜仁路）和爱文义路（今北京西路）恰好蕴含了主人的名字，而门牌 333 号据说系主人重金购得，代表占地三亩三分三厘。

该宅充分体现了有机建筑的设计原则。布局精密紧凑，与基地完美契合。主体紧贴北侧道路，与顺应转弯半径的弧形围墙连成整体。首层中间架空做汽车道，压缩面积。餐厅等和上层主人用房向花园开敞，北面的中式客堂、祖屋和用人房较封闭。半圆形楼梯外通高的玻璃正对道路转角：南立面设计精湛。圆柱形的阳光房通高四层，与层层退进曲线流畅的大露台形成纵横对比，顺餐厅外墙盘旋而上的弧形大楼梯将露台与花园连为一体。楼梯与阳台的铸铁花饰为艺术装饰风格，阳光房外均采用进口的圆弧玻璃，甚至连玻璃移门也是弧形的。

住宅室内装修豪华。日光室上覆玻璃顶棚，小舞厅安装弹簧地板，煤卫冷暖设备齐全，并首次在上海私宅安装了电梯，还是独特的荷叶形平面。车道两侧和楼梯墙面采用意大利洞石，墙面铸铁花饰为艺术装饰风格（今已不存）。一层的佛堂和祖屋则完全是中式风格。

1949 年后，该楼一二层被捐作上海工商联活动俱乐部。1978 年后，这里划归上海市规划设计研究院做办公楼，后改为晒图室。

如今，“绿房子”部分空间成为“规划师之家”和“上海城市规划博物馆”。

On June 14, 2014, China's ninth Cultural and Natural Heritage Day, there was a long queue of more than 300 meters around 333 Tongren Road although the temperature in Shanghai reached more than 30 degrees in the morning. In only three hours, more than 1,200 people flocked into the renovated "Green House" to get a glimpse of its new look. As Czech-born architect Ladislav Hudec once boasted to the owner at that time, "The residence I designed for you will not be outdated even after 100 years."

The Former Residence of D.V. Woo is Hudec's masterpiece in his prime time of career. It is a classic among modern garden houses, with a building area of nearly 1,700 square meters. As the façade and surrounding walls are all covered with green glazed tiles, it is commonly known as the "Green House." It is announced as Heritage Architecture in Shanghai.

D.V. Woo was a famous dye tycoon in old Shanghai. D.V. Woo made his fortune out of the green pigments for military use, so he regarded green as his lucky colour. The plot was his wife's dowry. The Chinese names of the adjoining Hardoon Road ("哈同路" in written chinese, today Tongren Road) and Avenue Road ("爱文义路" in written chinese,today West Beijing Road) happened to contain the two characters in the owner's name (同文). And the door plate 333 was said to be purchased by the owner with big money, representing the building took up an area of 3.33 mu.

The house fully embodies the design principle of organic architecture: the layout is precise and compact, and it balanced perfectly with the natural surroundings. The main building is close to the road on the north side and connected with the arc-shaped surrounding walls which conforms to the turning radius. The middle of the ground floor is built as a driveway to reduce the area. The dining room and the master's rooms upstairs are open to the garden, while the Chinese style guest hall, ancestral hall and servant rooms in the north are relatively closed. The glass outside the semicircular staircase is facing the turning of the road.

The south façade is exquisite in design: the cylindrical sunroom is four stories high, which forms a vertical contrast with the large horizontal terrace in smooth curves; and the large arc staircase circling along the dining room wall connects the terrace and the garden. The cast iron ornamental design of stairs and balconies is of Art Deco style. Imported arc glass is used to build the sunroom, and even the glass sliding door is arc-shaped.

The interior decoration is luxurious: the sunroom has a glass ceiling; the small ballroom is equipped with spring floors, and the whole house is equipped with complete cooling and heating facilities. It is the first private house in Shanghai to install an elevator, and the elevator car's floor is in the shape of a lotus leaf. Either side of the driveway and the staircase wall are surfaced with Italian travertine, and the cast iron ornamental design on the wall surface is of the Art Deco style (no longer exists today). The Buddhist Prayer Room and the ancestral hall on the first floor are completely of Chinese style.

After 1949, the first and second floors of the building were donated to the Shanghai Federation of Industry and Commerce and used as its club. After 1978, it was allocated to Shanghai Urban Planning & Design Research Institute as an office building, and later it was changed into a place to blueprint technical drawings.

Today, part of the "Green House" has been converted into "Home for Urban Planners" and "Shanghai City Planning Museum".

# 马勒别墅

Moller Villa

陕西南路 30 号有一幢以褐色耐火砖墙为主体的北欧风格建筑，在上海的建筑群中，因其别具一格而享有盛名。

1859 年，英籍犹太人赉赐・马勒在上海创办了“赉赐洋行”，随后成为上海滩的富商。1911 年，老马勒正式退休，长子伊利克・马勒接管业务。

1922 年，伊利克・马勒购进了福熙路（今延安中路）南侧亚尔培路（今陕西南路）的土地，建造私人住宅。他的弟弟林德赛・马勒是工程师，同时也是建筑设计爱好者，参与了住宅的设计。马勒别墅总体上属于北欧风格建筑，坡屋顶上增加许多尖屋顶，尖屋顶加开“老虎窗”，增加室内的采光。也许由于马勒家族是以航海起家的，而林德赛・马勒本身就是一个船舶修造工程师，于是室内装修多了许多邮船的元素，有多处形似舷窗的窗户，在扶梯的天花上大量采用如船舵等船用构件图案作装饰。在住宅的西侧三楼有一房间，设计了一长 2.5 米、宽约一米多的齐腰椭圆形围栏，栏内的地板像两扇门，可以翻启，当“门”启开时，站在栏杆旁一直可以看到底层的房间，许多人一直对这“家具”无法理解，实际在船上工作过的人一看就明白，轮船的机房一般设计在船尾的底舱里，在甲板通往机房处，就可以看到这种扶栏，站在此处可以看清机房的工作情况，也可以直接与机房操作工人对话。

1949 年后，马勒别墅成为共青团上海市委机关所在地。2002 年，归衡山集团所有，现作为宾馆对外开放。2006 年，被公布为全国重点文物保护单位。

There is a Nordic style building with brown fire-resistant brick wall at 30 South Shaanxi Road, known far and wide for its unique style among building complexes in Shanghai. There is a very magical and beautiful legend about the house: The owner had a beloved little daughter. She dreamed of entering a fairyland. After waking up, she drew a picture of the so-called fairyland, which became the blueprint for the architectural design of the villa.

In 1859, Nils Moller, a British Jew founded "Moller & Co., Ltd. (Shanghai)" in Shanghai on his own. He became a wealthy merchant in Shanghai. In 1911, the senior Moller officially retired and his first son Eric Moller took over the family business. The company gradually gave up the operation and agency trade business, and made every effort to develop the shipping industry. Therefore, he founded Moller Shipyard Co., Ltd. in Pudong, which had more than 2,000 employees in its heydays, and that was the predecessor of today's Shanghai Hudong Shipyard.

In 1922, Eric Moller bought land on Avenue du Roi Albert (today's South Shaanxi Road), south of Rue Foch (today's Middle Yan'an Road) to build a private house. His younger brother Lindsay Moller was an engineer and an architectural enthusiast. He supervised the shipyard and participated in the design of the villa. In general, the Moller Villa belongs to the Nordic-style architecture. Many peaked roofs are added to the slope roof and built with dormers to increase the indoor lighting. Perhaps because the Moller family made their fortune in shipping, and Lindsay Moller was an engineer in shipbuilding, the interior decoration is embellished with many ship elements. Many windows are in the shape of portholes. On the ceiling of the escalator, a large number of ship components such as rudder are used for ornaments. There is a room on the third floor of the west wing of the house. A waist-deep oval railing with a length of 2.5 metres and a width of over 1 metre is designed. The floor inside the railing is like two door leaves, which can be lifted up. When the "door" is opened, one can directly see the rooms on the ground floor at the railing. Many people found it incredible but it is quiet easy for seasoned sailors to understand. Usually the engine room of a ship is designed at the bilged compartment. And this kind of railing can be seen on the deck leading to the engine room. Standing there, one can see the working condition of the engine room, and can also talk directly to the operators inside it.

After 1949, the Moller Villa became the seat of the Communist Youth League of the Shanghai Municipal Party Committee. In 2002, Shanghai Hengshan (Group) Corp. acquired it. Now it is open to the public as a hotel. In 2006, it was announced as a Major Historical and Cultural Site Protected at the National Level.

# 原沙逊别墅
Former Sassoon Villa

长宁区虹桥路2409号
2409 Hongqiao Road, Changning District

1910年，得到上海道的批准，在沪侨民购进西郊的230亩土地兴建虹桥高尔夫球场，位置就是今天的上海动物园。在远郊建球场，必须先筑一条从上海市区通往球场的马路，这条路穿过西郊的虹桥镇，就被叫做“虹桥路”。

新沙逊洋行的老板维克多·沙逊是个精明的商人，也是“房地产大王”。1930年，洋行以大中实业公司的名义购进了位于虹桥路底与“西郊高尔夫球场”相邻的土地建造别墅，沙逊给这幢别墅取名“Eden Garden”。“Eden”通常汉译为“伊甸园”，常被用于比喻远离城市喧嚣的安乐之地。

沙逊别墅的主建筑位于花园的中间，主建筑以朝南的立面为主立面，典型的英国乡村建筑风，坡度很陡的红瓦大屋顶几乎占据了立面的四分之三的视线，红瓦绿地，互为呼应，使整幢建筑十分耀眼。沙逊还想方设法从英国进口原根的粗大橡木，建筑的下部以此为墙，经过精心设计的粗大木构架显露于外，使人产生一种回归乡村、回归自然的真切感。当然，他也不会忘记建筑的居住功能，室内装修更是不遗余力，木构架自然地暴露于外，墙面与之形成天然的组合，平添生活情调。

沙逊别墅于1989年被列为上海优秀近代建筑、上海市文物保护单位。

In 1910, with the approval of Shanghai Daotai, the foreign settlers in Shanghai purchased 230 mu of land in the western suburbs to build Hongqiao Golf Course at the spot today known as Shanghai Zoo. In order to build a golf course in the outer suburbs, a road from downtown leading to the course must be built first. As the road passed through Hongqiao Town in the western suburb, the road was named "Hongqiao Road".

Victor Sassoon, the boss of E. D. Sassoon & Co, was an astute businessman and a "Real Estate Mogul". In 1930, in the name of Dazhong Industrial Company, the firm purchased the land adjacent to the "Western Suburb Golf Course" at the end of Hongqiao Road to build a villa. It was a peaceful land far away from the hustle and bustle of the city, and Sassoon named it "Eden Garden".

Located in the middle of the garden, the main building of Sassoon's Villa with its façade facing south is a typical English countryside villa. The red-tiled roof with very steep slope takes up nearly three-quarters of the façade. With red tiles in striking contrast to the green garden, it is very eye-catching. Sassoon also managed to import thick oak logs from England to build the walls in the lower part of the building. The carefully designed thick wood frame was exposed, giving people a feeling of the rustic countryside and the nature. Of course, he took the residential function of the building in consideration as well. The interior decoration was even more extravagant. The wood frame was naturally exposed in a style with the wall surface, giving the villa an idyllic touch.

Sassoon's Villa was listed as a Monument under the Protection of the Shanghai Municipality in 1989.

# 上海汾阳花园酒店

Shanghai Fenyang Garden Hotel

徐汇区汾阳路45号
45 Fenyang Road, Xuhui District

上海音乐学院正门的对面，有一幢西班牙风格的假三层砖木石混合结构建筑，现在以“丁贵堂旧居”名义被公布为上海市文物保护单位。

1914 年，江海关从一位侨民手中购进一块位于毕勋路（今汾阳路 9 号）的土地，连同土地上已经建造的一幢平房，作为驻上海的副总税务司住宅。1930 年，江海北关又在这块土地的西南部建造副总税务司新宅。新住宅总占地约七亩，建筑面积 1236 平方米，为典型的西班牙风格建筑，假三层砖木石混合结构，坐北朝南，以南立面为主立面，以中间为中轴线，两侧不严格对称，中间底层设计为三连续拱支撑的内廊，由台阶相接，与客厅相通，二层退而为平台，三层即假三层，开设较大的屋顶窗，东西两侧的底层均为西班牙建筑风格的“帕拉第奥式窗”，这种窗特征明显，就是中间是一带丰圆拱窗楣的大窗，两边是高度相同、明显收窄的平顶小窗，中间的大窗与两边的小窗以螺旋纹饰的小柱相连。

1953 年，汾阳路 45 号成为上海海关学校。1980 年，上海海关学校改为上海海关专科学校，后迁往浦东新校址，这里仍由海关总署管理和使用。2010 年，汾阳花园酒店在这里开业。

On the opposite side of the main entrance of the Shanghai Conservatory of Music is a false three-storey masonry structure in Spanish style. It is declared as protected monuments of Shanghai under the name of "Former Residence of Ding Guitang".

In April 1914, the Shanghai Customs purchased a plot located on Route Pichon (today's 9 Fenyang Road) from a foreigner living in Shanghai together with a bungalow already built on the plot as the residence of the Deputy Inspector General of the Chinese Customs in Shanghai. In 1930, the North Customs House built a new house for the Deputy Inspector General in the southwest of the plot. The new residence covers a land area of about 7 mu and a floor area of 1,236 square metres, showing typical Spanish style. It is a false three-storey masonry structure facing south with the south façade as the main façade and the middle as the central axis, though the two sides are not strictly symmetrical. The middle of the ground floor is designed as a middle corridor supported by three arches in a row, connected with the living room by stairs and retreats back as a platform on the second floor. The third floor is the false storey where large roof windows are built. The ground floor on the east and west sides are the "Palladian" windows of Spanish architectural style. Such windows have distinctive characteristics, featuring a large window with circular arch lintel and two small flat top narrowed windows of the same height on both sides, and the three are connected by columns carved with spiral patterns.

In 1953, 45 Fenyang Road was built as Shanghai Customs School. In 1980, Shanghai Customs School was changed into Shanghai Customs College, and then relocated to the Pudong New Campus, and continued to be managed and used by China's General Administration of Customs. In 2010, Fenyang Garden Hotel opened for business here.

# 海上

Brilliant New Architecture

# 华章

# 上海展览中心

Shanghai Exhibition Centre

20 世纪五六十年代，不要说在南京路附近，几乎整个上海市区的人们都能在晚上准确地捕捉到一颗熠熠闪亮的红星——中苏友好大厦尖塔顶端的红色五角星。这处建筑后来更名为上海展览中心，作为 50 年代上海兴建的第一座大型建筑及第一个展览馆，在城市建设和发展史上有着重要的地位。

20 世纪 50 年代初，中央决定于 1955 年 3 月在上海举办“苏联经济及文化建设成就展”。经过考察与讨论，1954 年 4 月 10 日，中苏专家选定铜仁路哈同花园为展馆建设用地。双方经过日夜奋战，只用了 7 天时间，就奇迹般地完成了设计方案。5 月 1 日，开工典礼举行，5 月 4 日，正式开工。

值得一提的是，上海是软土地基，造高楼都要打桩。苏联专家则认为打桩对于上海土壤没有多大用处，只要使结构物具有足够的强度与稳定性，抵抗不均匀的沉降就可以了。因此，建造中苏友好大厦没有打一根桩。在中央大厅的基础工程中，采用了当时还比较新颖的结构概念“箱型基础”——将土地平整后，用钢筋混凝土浇捣出一整块的大箱底，再在上面盖楼，上下一体，要沉降就一起沉降。建设者们战胜了一个又一个困难，1955 年 3 月初工程按时完工。3 月 15 日，落成庆典在中苏友好大厦中央广场上隆重举行。3 月 16 日，“苏联经济和文化建设成就展览会”在大厦开幕。自此，中苏友好大厦成为上海的标志性建筑。

展览中心整个楼群呈现俄罗斯古典主义建筑风格，局部糅合了巴洛克艺术特点。主体建筑由序馆、中央大厅和电影院组成，前方东西两翼是二层展览馆，由一层围廊围合成庭院。主体建筑与东西围廊环抱的中央广场面积达 8000 平方米，中央有一座大型喷水池。塔楼上矗立镏金钢塔，塔顶上镶有直径四米的五角星，总高 110.4 米。

1968 年，中苏友好大厦改名为上海展览馆，1984 年改为现名。1989 年被评为“上海十佳建筑”，1999 年又被评为“新中国上海十大金奖经典建筑”。2005 年，被评为上海市第四批优秀历史建筑。至今仍是举办各种大型展览的理想场所，是上海主要的城市景观和标志性建筑之一。

In the 1950s and 60s, people in Shanghai, not only those living near the Nanjing Road, but those in the whole downtown area, could see clearly a shining red star at night, which topped the spire of the Sino-Soviet Friendship Building. Later renamed the Shanghai Exhibition Centre, the building has played an important role in the history of the city's construction and development for being the city's first large building and first exhibition hall built in the 1950s.

In the early 1950s, the Central Committee of the CPC decided to hold an exhibition to showcase the Soviet Union's economic and cultural achievements in Shanghai in March 1955. After investigation and discussion, Chinese and Soviet experts chose the Hardoon Garden on Tongren Road as the construction site on April 10, 1954. The two sides worked day and night and miraculously completed the design in just seven days. On May 1, the groundbreaking ceremony was held, and the construction began on May 4.

It is worth noting that as Shanghai had a soft soil foundation, piling was a must for building high-rises. Yet the Soviet experts believed that piling was not of much use and it could be abandoned as long as the structure was strong and stable enough to resist uneven settlement. Therefore, piling was skipped for the construction of the Sino-Soviet Friendship Building. In the foundation project of the central hall, a new structural concept of "box foundation" was adopted—after the land was leveled, a large box bottom was poured with reinforced concrete, on which the building was constructed. By doing so, the upper and the lower parts were integrated and would settle together. The builders overcame all the difficulties and completed the construction on time in early March 1955. On March 15, a grand inauguration ceremony was held in the central square. On March 16, the Exhibition on the Soviet Union's Economic and Cultural Achievements opened in the building. Since then, the Sino-Soviet Friendship Building has become a landmark in Shanghai.

The integrated building complex is in the Russian classical style, with part in the Baroque style. The main building is composed of a foyer, a central hall and a cinema. The east and west wings on the frontage are two-storey exhibition halls and the colonnades on the first floor form a courtyard. The central square, surrounded by the main building and the east and west colonnades, covers an area of 8,000 square metres, with a large fountain in the centre. A gilded steel spire, which stands on the tower, is topped by a star of four metres in diametre. The total height is 110.4 metres.

The Sino-Soviet Friendship Building was renamed the Shanghai Exhibition Hall in 1968, and the Shanghai Exhibition Centre in 1984. In 1989, it was honored as "Shanghai's Top Ten Buildings", and in 1999, as "Shanghai's Top Ten Golden-Award Classical Buildings in the New China". In 2005, it was listed among the fourth batch of outstanding historic buildings in Shanghai. As one of the city's major attractions and landmarks, the centre remains an ideal venue for holding grand exhibitions.

东方明珠广播电视塔
Shanghai Oriental Pearl Radio & TV Tower

浦东新区世纪大道1号
1 Century Avenue, Pudong New Area

在黄浦江边的陆家嘴，耀眼的东方明珠广播电视塔是上海最著名、最热闹的景观之一。这里原是滨江的浦东公园，随着浦东开发开放，它作为这处热土上最早的重大工程项目，于 1991 年开始建设。

世界上的混凝土电视塔大多为单筒体，华东建筑设计院的设计方案则采用多筒结构。由三个直体圆筒、斜撑筒体和 11 个圆球组成立体框架体系。电视塔由太空舱、上球体、下球体、塔座和广场组成，93 米处为直径 50 米高四层的下球，272.5 米处为直径 45 米高九层的上球，350 米处为直径 16 米的太空舱，再上面是 118 米高的钢桅杆天线段。这样的设计抗震、抗风性能好，符合塔体的美观性和合理性。

除了担当广播电视塔的功能外，东方明珠还是一个集旅游、观光、娱乐、购物于一体的好去处。在 1995 年 5 月 1 日对外开放后，东方明珠成为无数国内外游客热切关注的观光景点。

东方明珠广播电视塔入选 2019 上海新十大地标建筑。今天，东方明珠电视塔作为上海标志性建筑是毋庸置疑的，其独特的造型已成为上海的一张城市文化名片。

Shanghai Oriental Pearl Radio & TV Tower, located at the tip of Lujiazui by the side of Huangpu River, is one of Shanghai's most famous and most-visited tourist attractions. Situated within what was originally the Pudong Park, Shanghai Oriental Pearl Radio & TV Tower broke ground in 1991 and was the first construction project to kick off in this land full of promise with the development and opening-up of Pudong.

Most concrete TV towers across the world tend to be single-tube structures, while the design of the East China Architectural Design & Research Institute adopted a multitube structure—a three-dimensional frame composed of 3 vertical stanchions, slanting stanchions and 11 spheres. Shanghai Oriental Pearl Radio & TV Tower is comprised of the space capsule, upper sphere, lower sphere, tower base and square. The lower sphere, with a diameter of 50 metres at 93 metres high, has four storeys, the upper sphere, with a diameter of 45 metres at 272.5 metres high, has nine storeys, and there is a space capsule with a diameter of 16 metres at 350 metres, above which is a 118-metre-high steel antenna mast. It can resist earthquakes and strong wind, and its appropriately designed body creates an amazing view.

Apart from serving as a radio and television tower, it is also a go-to place for tourism, sightseeing, entertainment and shopping. Since it was put into use on May 1, 1995, Shanghai Oriental Pearl Radio & TV Tower has become a must-see sight of Shanghai, attracting countless domestic and foreign tourists.

Shanghai Oriental Pearl Radio & TV Tower was declared one of 2019's Top 10 Landmark Buildings in Shanghai. As a well-deserved landmark of Shanghai, it has become a cultural symbol of the city with its unique design.

## 上海大剧院
Shanghai Grand Theatre

黄浦区人民大道300号
300 Renmin Avenue, Huangpu District

在上海人民大道西北端，于 1994 年开始建造上海大剧院。大剧院由法国夏氏建筑设计事务所设计方案，华东建筑设计研究院进行施工图设计。建筑占地面积 1.1 万平方米，建筑面积 6.2 万平方米，地上八层，地下两层，总高度 40.5 米。

建筑由 1800 座大剧场、600 座中剧场和 300 座小剧场组成，可用于歌剧、芭蕾舞和交响乐演出。大剧场舞台由主舞台、后台和两边副台组成，主舞台可整体提升，也可分块提升，侧面副台可平移至主舞台，后台为旋转舞台还可旋转至主舞台。舞台前的 100 平方米乐池，也可自动升降。观众厅有三层，底层从前排至后排坡高五米，使观众获得最佳视线。二层、三层楼座和侧面六个包厢环抱布置，将观众厅华丽呈现。

建筑立面方正，南立面大台阶拾级而上，宽敞的大堂面积有 2000 平方米，高 18 米，内部装饰白色基调。大堂悬挂六片排箫灯架组合而成的大水晶吊灯，地面铺白色大理石，四周是钢索玻璃幕墙，用钢索张拉结构支撑。在灯光下，大堂晶莹透亮，犹如“水晶宫”，将内部结构等暴露出来，极具科技感。按照中国古典建筑亭的外形设计，屋顶采用两边反翘和天空拥抱的白色弧形，寓意天圆地方之说，也象征各国灿烂文化的巨大“聚宝盆”。

自 1998 年 8 月开业以来，上海大剧院一直致力于实现“中国剧院标杆、城市文化名片、文化创意中心”的发展愿景。

The Shanghai Grand Theatre project started in 1994 at the northwest end of Ren Min Avenue. Arte Charpentier was responsible for the design whilst East China Architectural Design & Research Institute for the construction drawings. The 40.5-metre-high building is spread over 11,000 square metres with a gross floor area of 62,000 square metres, divided into eight floors above ground and two floors below ground.

The development consists of a 1,800-seat Lyric Theatre, a 600-seat Buick Theatre and a 300-seat Studio Theatre, to accommodate performances including opera, ballet and symphonic music. The avant-garde stage of Lyric Theatre is made up of a main platform, a rear revolve and wings, all removable in different layers as plots need, in front of which there is a 100-square-metre automatic orchestra pit. To ensure a best view, the magnificent auditorium is on three levels with the stalls rising away from the front on a 5-metre slope, balconies on dress circle and upper circle, and six boxes on both sides.

The building has a square facade, with a grand staircase at the main entrance on the south. White-toned in general, its lobby is as spacious as 2,000 square metres and is 18 metres in height, with a six-pan pipe shaped chandelier hanging from above, a floor made up of white marble and a glass curtain wall supported by a wire tension structure. When the night falls and the light goes on, the building turns into a crystal palace, where the structure of the lobby transcends with a sense of high-tech art. Inspired by the shape of Chinese pavilions, the upper part of the building is concave, embracing the sky. It is a symbol of an ancient Chinese perception of the world as "round sky, square earth" and of a "treasure bowl" mixing splendid cultures of various countries as well.

Since its opening in August 1998, The Shanghai Grand Theatre has been committed to the vision of "the benchmark of Chinese theatres, the name card of urban culture and the creative center of culture".

# 金茂大厦

Jinmao Tower

今时今日，站在花园石桥路，以路为界，举目四望，其北、东、南三边，分别矗立着金茂大厦、上海环球金融中心和上海中心大厦。陆家嘴上，三足鼎立的建筑，标志着上海高楼的高度，光辉夺目。其中，金茂大厦是这三座超高层建筑中最早建成的。

从旧日低矮传统民居区到变身成为高楼区，这份拔地而起的胆识和速度，只有奇迹二字可以概括。而这，也是浦东开发开放速度的直观写照。

1992 年，国家经贸部决定在上海浦东陆家嘴金融区建造“第一高度”的建筑，1999 年 3 月，88 层高的金茂大厦投入使用。塔楼造型从中国古代宝塔建筑上获得设计灵感，不锈钢线条与玻璃幕墙结合，层层叠叠向上伸展，直至高耸的塔尖。裙房立面由花岗石与不锈钢结合，显示坚固和稳重。白天塔楼金属表皮银光闪闪，晚上犹如发光的灯塔。

在当时中国第一、世界第三的金茂大厦内，除了购物中心、办公区域和会展区域外，53—87 层为超五星级宾馆。旅客从底层裙房的宾馆入口处，可以乘六座高速电梯到达宾馆空中大堂，中央有 30 层高 152 米、直径 27 米的中庭，28 道环行灯廊层层叠叠，仿佛“时空隧道”，雄伟壮观。88 层的观光厅为 360 度环形大厅，可俯瞰浦江两岸及市中心美景。

Standing on Huayuan Shiqiao Road and looking around, you will see the Jinmao Tower, Shanghai World Financial Center and Shanghai Tower respectively erect on the north, east and south sides of the road. The three eye-catching buildings at Lujiazui are the highest among Shanghai's buildings. Among them, the Jinmao Tower was the first skyscraper completed.

The area of high-rise buildings are transformed from that of the old low-rise traditional residential buildings with amazing boldness and speed, which can only be described as a miracle. This is also a visual portrayal of Pudong's fast development and opening.

In 1992, the then Ministry of Economy and Trade (now the Ministry of Commerce) decided to build the highest building in Lujiazui, Shanghai. In March 1999, the 88-storeyed Jinmao Tower was put into use. Inspired by the ancient Chinese pagoda in its appearance design, the Tower combines stainless steel with glass curtain walls that stretch upward layer after layer until the spire. The podium facade is composed of granite and stainless steel, which are robust and stable. Glimmering with its metal appearance during the day, the tower looks like a glowing lighthouse at night.

The Jinmao Tower, the highest in China and the third highest in the world upon its completion, accommodates a super five-star hotel from the 53rd to 87th floor in addition to the shopping mall, office area and exhibition area. Visitors can take six high-speed elevators at the hotel entrance of the ground floor podium to reach the sky lobby. There is also a 152-metre-high and 27-metre-diametre atrium in the centre, and 28 circular light corridors stacking one upon another, as magnificent as a "space-time tunnel". The 360-degree ring-shaped sightseeing hall on the 88th floor overlooks the banks of the Huangpu River and city centre.

# 上海世博会中国国家馆、世博文化中心
Shanghai World Expo China Pavilion and World Expo Cultural Center

浦东新区上南路205号，世博大道1200号
205 Shangnan Road, 1200 Expo Avenue, Pudong New Area

上海浦东世博园区南北、东西轴线交会处的核心地段，坐落着中国国家馆。中国馆大台阶共有 76 级，用“紫晶白麻”和“华夏灰”两种石材筑成，这台阶表面的纹理，是一批顶尖石匠用一种濒临失传的绝技“三斩斧”纯手工“斩”成，比国外顶级机器加工出来的纹理更加均匀和细致。

沿着这台阶往上走去，便是被称为“东方之冠”的中国国家馆作为 2010 年上海世博会期间人人争睹的头号明星。

这一世博会核心建筑，从 344 件应征设计方案中脱颖而出，由建筑设计师何镜堂主持，综合华南理工大学设计院、清华安地建筑设计公司和上海建筑设计研究院的方案，于 2010 年 2 月 8 日竣工，在世博会后被永久保留，现为中华艺术宫。

中国国家馆设计体现“城市，让生活更美好”主题，展示城市发展中的中华愿景。地方馆设计提供给 31 个省、直辖市和自治区展示各民族风采及建设成就。为此，建筑造型居中突起，形成冠盖，层叠出挑，制似斗栱，四根大柱托起，外观“东方之冠”的构思象征中华崛起。斗栱内 56 根横梁叠加，象征 56 个民族团结的力量。地方馆紧紧围合中央形成稳定的基座，屋顶平台达 2.7 万平方米，布置来自圆明园“九洲清晏”，寓意“田”“泽”“渔”“脊”“林”“甸”“壑”“漠”景观，既能供人流集散，又可供休闲。整个中国馆立面颜色以北京的故宫“红”为基调，大气恢宏。

世博轴东侧，是被昵称为“飞碟”的世博文化中心，为园区最大演出场地。作为国内第一个容量可变的大型室内场馆，其剧场空间可以根据需要，隔成 18000 座、12000 座、10000 座、8000 座、5000 座等。

2010 年 10 月 31 日晚 8 点，上百名少年儿童手持绚丽的花束和花环走上上海世博文化中心舞台中央，中国 2010 年上海世界博览会圆满闭幕。次日起，世博文化中心更名为梅赛德斯-奔驰文化中心。

世博文化中心
World Expo Cultural Center

The China Pavilion sits at the intersection of the north-south and east-west axes of the Shanghai Expo Park. It has 76 big steps made of two kinds of granites — "amethyst and seasame white granite" and "Chinese gray granite". The patterns on the surface of the steps are carved by the best stonemasons with axes using a nearly lost skill called "three chop", which are more even and delicate than those processed by the top foreign machines.

At the end of the steps lies the China Pavilion, known as the "Oriental Crown", which was the number one star that visitors scrambled to visit during the Expo 2010 Shanghai China.

The China Pavilion's final design plan stood out from 344 solicited ones. Presided over by the architect He Jingtang, it integrated the plans from the Architectural Design and Research Institute of South China University of Technology, Tsinghua Andi Architectural Design Company and Arcplus Institute of Shanghai Architectural Design & Research. The construction was completed on February 8, 2010. The China Pavilion was designed to be permanent and now becomes the China Art Museum, Shanghai.

The design of the China Pavilion embodies the theme of "Better City, Better Life" and showcases the national vision through urban development. The design of the Regional Pavilion aims to display the ethnic features of and construction achievements made by the 31 provinces, municipalities and autonomous regions. Therefore, the shape of China Pavilion protrudes in the middle, forming a crown cover layer upon layer, like a bucket arch propped by four large pillars. The "Oriental Crown" of its appearance symbolizes the rise of China. The superimposed 56 beams in the bucket arch represent the unity of China's 56 nationalities. The center area is tightly surrounded by the Regional Pavilion, which forms a stable base. The 27,000-square-metre roof platform draws inspiration from "Jiuzhou

Qingyan" in Yuanmingyuan or the Old Summer Palace, and represents the landscapes of "field", "lake", "fishing area", "ridge", "forest", "suburb", "valley" and "desert". It can be used not only for gathering and dispersing crowds, but also for visitors to have leisure. The China Pavilion's entire facade has a "red" tone, which is as magnificent as that of the Forbidden City in Beijing.

On the east side of the Expo Axis is the Shanghai World Expo Cultural Center nicknamed "flying saucer", which is the largest performance venue in the park. As the first large indoor venue with variable accommodation capacity in China, the theater facility can be divided into 18,000 seats, 12,000 seats, 10,000 seats, 8,000 seats, 5,000 seats, etc. as needed.

At 8:00 pm on October 31, 2010, hundreds of children walked to the center of the stage of the Shanghai World Expo Cultural Center with gorgeous bouquets and garlands, kicking off the closing ceremony of the Expo 2010 Shanghai China. The next day, the Center was renamed the Mercedes-Benz Arena.

## 上海中心大厦
Shanghai Tower

浦东新区陆家嘴环路479号
479 Lujiazui Ring Road, Pudong New Area

2016年4月，中国第一高楼上海中心竣工。在许多场合，它都被比作一个“垂直的城市”。

大厦总建筑面积57.8万平方米，相当于将外滩1.5公里沿岸的第一排近60万平方米建筑竖立起来。这座由9个垂直社区、127层楼组成的超级摩天大楼涵盖了五大功能：品牌商业服务、超甲级办公、超六星级酒店与精品办公、观光和文化娱乐以及大型活动和会议等。

拥有厚达280多米松软且含水量高的土壤的上海，不易建造超高层建筑。世界上尚无软土地基上建造600米以上超高层建筑的先例。经过严谨的实地试验和科学论证，工程师联合国内顶尖的桩基研究和施工专家，终于找到了一个解决办法——后注浆钻孔灌注桩，决定由1079根后注浆钻孔灌注桩，扛起主楼近百万吨的重量。

而楼上，内部的小世界同样精彩。上海中心能接纳的人数达到35000人，并有博物馆、园林等入驻。为了降低大风造成的大楼摇摆而安装的阻尼器因为利用“电磁原理”，更节省出了一片“上海慧眼”空间，海内外艺术家可在此举办小型演出或展览。

这座名副其实的上海之巅，在设计中采用了40多项绿色技术，如在大厦屋顶利用平均风速每秒8—10米风力，安装风力发电机；有效利用建筑雨水资源；LED 固态光源照明等，节能率可达约 54%，每年可减少碳排放量 2.5 万吨，是绿色环保先进建筑，获得多项国际或国家认证。

上海中心的 52 层开办了朵云书院旗舰店，拥有绝佳的空中视野和“抬头看云、低头看书”的阅读环境。239 米高处的旗舰店是目前中国最高的商业运营书店，成为上海重要的文化新地标。

上海中心为城市增添了新的天际线，入选 2019 上海新十大地标建筑。

In April 2016, the construction of Shanghai Tower, the tallest building in China, was completed. Shanghai Tower is often described as a “vertical city”.

With a floor area of 578,000 square metres, it is equivalent to the sum of the first-row buildings covering a total floor area of nearly 600,000 square metres along the 1.5-kilometre Bund. This 127-storey super skyscraper with nine vertical communities has five major functions, namely brand business services, super class A offices, super six-star hotels, sightseeing and culture & entertainment, as well as large-scale events & conferences.

It is a great challenge to build super high-rises in Shanghai, since its soft soil layer is more than 280 metres thick and has a high moisture level. Besides, there was no precedent in the world for the construction of ultra-high-rise buildings above 600 metres on soft soil foundations. After rigorous field tests and scientific demonstrations, the engineers of Shanghai Tower, together with leading experts in pile foundation research and construction in China, finally figured out a solution, a foundation system of post-grouted bored piles. They decided to use 1,079 post-grouted bored piles to prop up the main building weighing nearly 1 million tons.

Inside Shanghai Tower, which can accommodate 35,000 people, is a wonderful small world with museums, gardens, etc. A damper using the “electromagnetic principle” has been installed to reduce the impact of strong winds on the building, and inside the damper there is a space dubbed “Shanghai Eye”, where artists from home and abroad can present small performances or exhibitions.

Shanghai Tower, the highest building in Shanghai, adopts more than 40 green technologies in its design. For example, wind turbines are installed to make use of the wind force, with an average speed of 8-10 metres per second, on the roof; the rainwater on the buildings is collected and put into effective use; and LED solid-state lighting, with an energy saving rate of about 54 percent, can reduce carbon emission by 25,000 tons per year. This green and environmentally friendly building has won numerous international or national awards and certifications.

On the 52nd floor of Shanghai Tower is a flagship store of Duoyun Books, where visitors can "look up to watch the clouds wander by and look down to read books". Situated at a height of 239 metres, it is currently the highest commercial bookstore in China and an important new cultural landmark in Shanghai.

Shanghai Tower is a striking new addition to the skyline of the city; it was declared one of the 2019 Top 10 Landmark Buildings in Shanghai.

上海国家会展中心
National Exhibition and Convention Center (Shanghai)

青浦区崧泽大道333号
333 Songze Avenue, Qingpu District

上海国家会展中心是商务部和上海市人民政府于 2011 年合作共建的项目，是集展览、会议、活动、办公、宾馆于一体的超大型会展中心。选址在虹桥商务区核心段西部，临近虹桥机场和火车站，与地铁 2 号线徐泾东站对接，与城市高架和周边高速公路连通，由华东建筑设计研究院与清华大学设计院联合设计。

上海国家会展中心总建筑面积 147 万平方米，立面通过曲面幕墙和 240 根列柱，寓意“生命旺盛的草茎”，展现柔和飘逸又浑然大气的建筑形象。上海国家会展中心可展览面积达 50 万平方米，包括 40 万平方米的室内展厅和 10 万平方米的室外展场。

2018 年 11 月 1 日，第一届中国国际进口博览会开幕的前四天，长三角生态绿色一体化发展示范区正式揭牌，包含了进博会举办地“四叶草”的这片大区域。11 月 13 日，进博会向公众开放延展的第一天，上海公布《关于加快虹桥商务区建设　打造国际开放枢纽的实施方案》，同样包含了“四叶草”这片大区域，这如同一个个不同的注脚，人们一再读到“开放”之于上海的深意。

此后每年，“四叶草”都将举办进博会，成为中国向全世界开放的平台。人的流动，某种程度上也是实体展会的一大魅力所在。在互联网如此发达的当下，被称为“四叶草”的上海国家会展中心内，用空间展现了一份线上世界不可能取代的交流奇迹：构造一个特别的、具有交往意义的平台。作为中国着眼于推动新一轮高水平对外开放作出的重大决策和中国主动向世界开放市场的重大举措，以“四叶草”为标志的进博会见证着中国的开放和上海日新月异的发展。

NECC (Shanghai) is a cooperation project launched by the Chinese Ministry of Commerce and the Shanghai Municipal People's Government in 2011, designed to be a super center for holding exhibitions, conferences and events and offering office space and hotel services. It is located in the west of the core section of Hongqiao Central Business District, close to Hongqiao Airport and Hongqiao Railway Station, connected to the East Xujing Station of Shanghai Metro Line 2 as well as the city's elevated roads and nearby expressways. It was jointly designed by the East China Architectural Design & Research Institute (ECADI) and the Architectural Design & Research Institute of Tsinghua University.

With a total floor area of 1.47 million square metres, NECC (Shanghai) has a façade featuring curved curtain walls and 240 pillars, which symbolize "the exceptional vitality of grass" and present a sleek yet magnificent image. It has a total exhibition area of 500,000 square metres, including 400,000 square metres indoors and 100,000 square metres outdoors.

On November 1, 2018, four days before the opening of the first CIIE, the demonstration zone for integrated green development of the Yangtze River Delta region was officially unveiled. The zone includes the "four-leaf clover" where the Expo would be held. On November 13, the first day that the Expo opened to the public, Shanghai announced the Implementation Plan on Accelerating the Development of Hongqiao Central Business District and Creating an Open International Hub. The "four-leaf clover" was once again included in the plan. These are just like footnotes to the significance of "openness" to Shanghai.

Every year after 2018, the "four-leaf clover" will stage the CIIE and serve as China's open platform to the world. Part of what is fantastic about physical exhibitions is the flow of visitors. Despite the Internet being so advanced now, NECC (Shanghai), nicknamed "four-leaf clover", uses space to build a special platform that facilitates interactions, a miracle for exchanges that cannot be replicated in the online world. As an important decision made by China to pursue a new round of high-level opening-up and China's major initiative to still widen market access to the rest of the world, with the "four-leaf clover" as its symbol, CIIE has witnessed China's opening-up and Shanghai's rapid development.

# 后记

2019 年,《这里是上海：建筑可阅读》系列出版计划启动，包括精装本、普及本、有声版等多种形态，由上海市文化和旅游局（市文物局）作为组织协调单位，上海城市推广中心作为出品单位，与上海世纪出版集团联合推出。本书由宗明担任主编，编委会成员包括于秀芬、王为松、包亚明、伍江、汤惟杰、孙甘露、李天纲、张伟、陈丹燕、陈保平、郑时龄、俞敏亮、姜澜、姚映然、顾洪辉、徐锦江、唐玉恩、黄强、曹嘉明、常青、褚晓波、阚宁辉、谭玉峰、薛理勇（按姓氏笔画排序）。

本书的编辑出版，得到本市有关单位、部门的关心指导，以及相关专家、学者的支持参与：郑时龄、常青、伍江、谭玉峰等对相关建筑内容予以审定；王海松、左琰、朱少伟、朱珉连、朱晓明、乔争月、华霞虹、汤惟杰、吴海勇、吴皎、张伟、陈丹燕、陈伟、金波、周进、郑瑛、娄承浩、祝东海、顾定海、曹永康、曹伟明、薛理勇（按姓氏笔画为序）等撰写相关篇目；张伟、汤惟杰、沈铁伦梳理润色书稿文字；陈保平、李天纲审读全书内容；刘开明、顾铮对图片提出专业意见；中译语通信息科技（上海）有限公司承担英文翻译；上海日报社、英汉大词典编纂处，Andy Boreham、Emma Keeley Leaning、Lancy Correa、刘琦、张颖、王梓诚、吕栋等审校英文书稿；上海市文化和旅游局（市文物局）、各区文化和旅游局、上海市司法局、文汇报社、上海孙中山故居纪念馆、上海宋庆龄故居纪念馆、美琪大戏院、上海大剧院、上海东方明珠广播电视塔、上海中心、上海红砖文化传播有限公司和刘文毅、齐琦、汤涛、许一凡、孙爱民、吴禹星、沈平允、张旭东、周文强、袁婧、郭长耀、席闻雷、陶钧、黄伟国（按姓氏笔画为序）等单位与个人提供图片支持；徐威、朱家健、欧晓川、袁斌、吴海勇等对相关书稿内容给予协调帮助。

谨向在本书编辑出版过程中提供支持的单位和个人，致以诚挚的谢意。

上海人民出版社

2020 年 7 月

# Afterword

In 2019, Stories of Shanghai Architecture publishing plan started, including hardback, paperback, audio version and other forms.It is launched by Shanghai Century Publishing Group, sponsored and coordinated by the Shanghai Municipal Administration of Culture and Tourism (Shanghai Municipal Administration of Cultural Heritage), and produced by Shanghai City Promotion Office. The editor-in-chief is Zong Ming, the editorial board includes Yu Xiufen, Wang Weisong, Bao Yaming, Wu Jiang, Tang Weijie, Sun Ganlu, Li Tiangang, Zhang Wei, Chen Danyan, Chen Baoping, Zheng Shiling, Yu Minliang, Jiang Lan, Yao Yingran, Gu Honghui, Xu Jinjiang, Tang Yu'en, Huang Qiang, Cao Jiaming, Chang Qing, Chu Xiaobo, Kan Ninghui, Tan Yufeng, Xue Liyong(stoke order of surname).

The project has received the care and guidance from related organizations and authorities of the Shanghai Municipality as well as the support and participation of experts and scholars. Prof. Zheng Shiling, Prof. Chang Qing, Prof. Wu Jiang, Prof. Tan Yufeng and other experts were specially invited to review and approve the content of the architecture in the book. Wang Haisong, Zuo Yan, Zhu Shaowei, Zhu Minwu, Zhu Xiaoming, Qiao Zhengyue, Hua Xiahong, Tang Weijie, Wu Haiyong, Wu Jiao, Zhang Wei, Chen Danyan, Chen Wei, Jin Bo, Zhou Jin, Zheng Ying, Lou Chenghao, Zhu Donghai, Gu Dinghai, Cao

Yongkang, Cao Weiming, and Xue Liyong (stoke order of surname) wrote the text. Zhang Wei, Tang Weijie and Shen Yilun combined and polished the manuscripts. Chen Baoping and Li Tiangang proof-read and copy-edited the book. Liu Kaiming and Gu Zheng offered professional advice for the courtesy of the photographs. Global Tone Communication Technology (Shanghai) Co., Ltd. translated the book from Chinese to English. Shanghai Daily, Editorial Office of the English-Chinese Dictionary, Andy Boreham, Emma Keeley Leaning, Lancy Correa, Liu Qi, Zhang Ying, Wang Zicheng and Lv Dong are the English proof-readers. We also highly appreciate the photographs offered by organizations and individuals, including the Shanghai Municipal Administration of Culture and Tourism (Shanghai Municipal Administration of Cultural Heritage), Municipal Administration of Culture and Tourism in each district, the Shanghai Municipal Bureau of Justice, Wen Hui Bao, Memorial Hall of Dr. Sun Yat-sen's Former Residence in Shanghai, Soong Ching Ling Memorial Residence in Shanghai, Grand Theatre, Shanghai Grand Theatre, Shanghai Oriental Pearl Radio & TV Tower, Shanghai Tower, Shanghai D.P. Company Ltd. as well as Liu Wenyi, Qi Qi, Tang Tao, Xu Yifan, Sun Aiming, Wu Yuxing, Shen Pingyun, Zhang Xudong, Zhou Wenqiang, Yuan Jing, Guo Changyao, Xi Wenlei, Tao Jun, and Huang Weiguo (stoke order of surname). And the related coordinate and help of the manuscript content from Xu Wei, Zhu Jiajian, Ou Xiaochuan, Yuan Bin and Wu Haiyong.

We would like to extend our sincere thanks to all the organizations and individuals who have rendered support in editing and publishing this book.

Shanghai People's Publishing House

July 2020

守 望 思 想　　逐 光 启 航

**这里是上海：建筑可阅读**

宗　明 主编

责任编辑　肖　峰
营销编辑　池　淼　赵宇迪
英文翻译　中译语通信息科技（上海）有限公司
封面设计　周伟伟
版式设计　徐　翔

出版：上海光启书局有限公司
地址：上海市闵行区号景路 159 弄 C 座 2 楼 201 室　201101
发行：上海人民出版社发行中心
印刷：浙江经纬印业股份有限公司
制版：南京理工出版信息技术有限公司

开本：787mm × 1092mm　1/32
印张：11　字数：156,000
2023 年 9 月第 1 版　2024 年 8 月第 2 次印刷
定价：110.00 元
ISBN：978-7-5452-1984-5 / T · 2

**图书在版编目(CIP)数据**

这里是上海 : 建筑可阅读 / 宗明主编 . —上海 :
光启书局，2023（2024.8 重印）
ISBN 978-7-5452-1984-5

Ⅰ . ① 这…　Ⅱ . ① 宗…　Ⅲ . ① 建筑艺术－介绍－上海
Ⅳ . ① TU-862

中国国家版本馆 CIP 数据核字 (2023) 第 161796 号

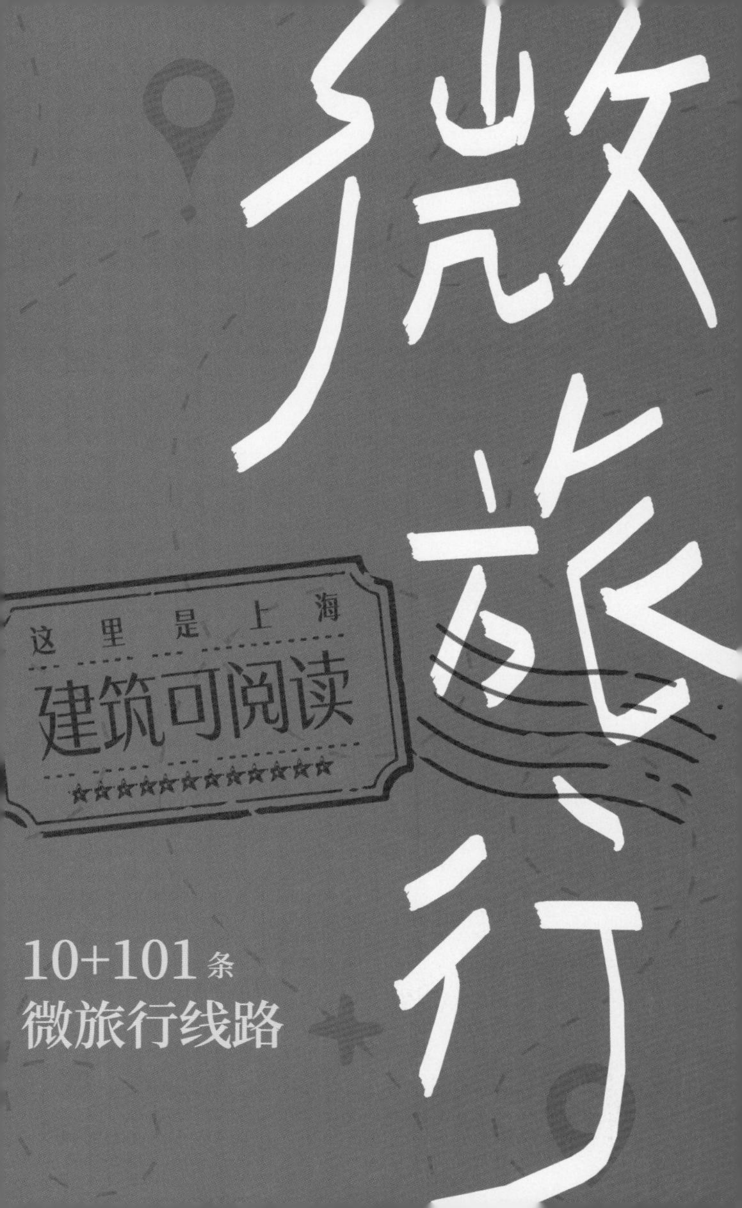

微旅行
这里是上海
建筑可阅读
10+101条
微旅行线路

用脚步丈量上海，
聆听历史建筑前世今生，
感受城市历史人文气息。

# #1 难忘红色记忆

**打卡沪上经典红色景点，感受红色文化的历史与发展，**
**重温党的奋斗历程，凝心聚力共创未来。**

**交通方式：**
步行、公交

**打卡点：**
中共一大会址纪念馆（黄浦区兴业路 102—108 号）→周公馆（黄浦区思南路 73 号）→孙中山故居（黄浦区香山路 7 号）→渔阳里（黄浦区淮海中路 567 弄 1—23 号）→中共二大会址纪念馆（静安区老成都北路 7 弄 30 号）→毛泽东旧居（静安区茂名北路 120 弄 7 号）→上海市历史博物馆 & 上海革命历史博物馆（黄浦区南京西路 325 号）

# #2 感悟江河情怀

**苏州河与黄浦江，**
**上海文化生活与人文情感最为重要的空间载体，**
**在水畔感悟城市的情怀。**

**交通方式：**
步行、地铁或轮渡

**打卡点：**
上海邮政总局（虹口区北苏州路 250、258 号）→外滩建筑群 ( 黄浦区中山东一路 ) →东方明珠广播电视塔（浦东新区世纪大道 1 号）→金茂大厦（浦东新区世纪大道 88 号）→上海中心大厦（浦东新区陆家嘴银城中路 501 号）

# #3 重温休闲时光

**我们怀念过去优雅精致的摩登体验，**
**也将更加努力创造未来幸福美好的生活。**

**交通方式：**
步行、公交

**打卡点：**
上海音乐厅（黄浦区金陵中路 88 号）→大世界（黄浦区西藏南路 1 号）→原四大公司（黄浦区南京东路）→国际饭店（黄浦区南京西路 170 号）→大光明电影院（黄浦区南京西路 216 号）→美琪大戏院（静安区江宁路 66 号）

# #4 漫步梧桐深处

**高大的梧桐投下斑驳的树影，**
**各不相同的生活场景下，**
**藏着同样平凡和朴实的幸福。**

**交通方式：**

公交

**打卡点：**

步高里（黄浦区陕西南路与建国西路）→马勒别墅（静安区陕西南路30号）→裕华新村（静安区富民路182弄）→涌泉坊（静安区愚园路395弄4—24号）

# #5 寻踪名人传奇

**许多名人在上海留下了故事，**
**这一次的行走带你领略往昔的风起云涌**
**和曾经的惊世传奇。**

**交通方式：**
步行、公交

**打卡点：**
上海工艺美术博物馆（徐汇区汾阳路 79 号）→上海宋庆龄故居纪念馆（徐汇区淮海中路 1843 号）→武康大楼（徐汇区淮海中路 1850 号）→黄兴旧居（徐汇区武康路 393 号）→巴金故居（徐汇区武康路 113 号）→柯灵故居（徐汇区复兴西路 147 号）→张乐平故居（徐汇区五原路 288 弄 3 号）→蔡元培故居（静安区华山路 303 弄 16 号）→张爱玲故居（静安区常德路 195 号）→荣宗敬旧居（静安区陕西北路 186 号）

# #6 传奇建筑新生

**历史建筑与现代商业、现代文化的结合，**
**给城市带来了全新风景和生活方式。**

**交通方式：**
公交

**打卡点：**
“大上海计划”公共建筑群之绿瓦大楼（杨浦区长海路 399 号）→杨浦区图书馆（杨浦区长海路 366 号）→上海自来水科技馆（杨浦区杨树浦路 830 号）→ 1933 老场坊（虹口区溧阳路 611 号）→摩西会堂旧址（虹口区长阳路 62 号）→上海展览中心（静安区延安中路 1000 号）→上生・新所（长宁区延安西路 1262 号）

# #7 城市古迹印象

**繁华的都市里藏着这些古迹，**
**一石一木、一墙一殿，**
**真真实实穿越了千百年，遗世而独立。**

**交通方式：**

公交、地铁

**打卡点：**

大境阁（黄浦区大境路 259 号）→豫园（黄浦区豫园老街 279 号）→ 文庙（黄浦区文庙路 215 号）→三山会馆（黄浦区中山南路 1551 号）→龙华塔和龙华寺（徐汇区龙华路 2853 号）→桂林公园（徐汇区桂林路 128 号）

# #8 交汇中西文化

**海纳百川的上海，**
**中西方文化在这里交融、萌生，**
**铸就近代都市文明与艺术的先驱。**

**交通方式：**
公交、地铁

**打卡点：**
土山湾博物馆（徐汇区蒲汇塘路 55 号）→徐家汇天主堂（徐汇区蒲西路 158 号）→徐家汇藏书楼（徐汇区漕溪北路 80 号）→新华路外国弄堂→邬达克旧居（长宁区番禺路 129 号）→圣约翰大学旧址（长宁区万航渡路 1575 号）

# #9 畅游海上华章

**上海的每一幢新建筑，**
**都是改革开放的亲历者，**
**也是城市发展的见证者。**

**交通方式：**

步行、地铁

**打卡点：**

中华艺术宫（浦东新区上南路 205 号）→梅赛德斯 - 奔驰文化中心（浦东新区世博大道 1200 号）→东方体育中心（浦东新区泳耀路 300 号）→上海虹桥枢纽中心→上海国家会展中心（青浦区崧泽大道 333 号）（外观）

# #10 探访春申古风

郊游上海，探访那些承载历史记忆的古老建筑。

**交通方式：**
自驾

**A 线打卡点：**
嘉定孔庙（嘉定区南大街 183 号）→古猗园（嘉定区南翔镇沪宜公路 218 号）→醉白池（松江区人民南路 64 号）→古松江方塔园（松江区中山东路 235 号）→云间第一楼（松江区中山东路 250 号）
**B 线打卡点：**
佘山天文台（佘山国家森林公园内）→广富林文化遗址（松江区广富林路 3260 弄）→崧泽遗址博物馆（青浦区沪清平公路 3993 号）→金泽古镇普济桥（青浦区金泽镇）→曲水园（青浦区公园路 612 号）

# 黄浦区

#1 **点燃革命火种**

类型：红色文化
线路长度（km）：6
预计游览时间：半天

中国共产党第一次全国代表大会会址纪念馆→八路军驻沪办事处旧址→中国农工民主党第一次全国干部会议会址旧址→第一次国共合作时期上海执行部旧址→《新青年》编辑部旧址（中国共产党发起组成立地）→中国社会主义青年团中央机关旧址纪念馆→中共六大后党中央政治局机关旧址

#2 **感悟知识力量**

类型：红色文化
线路长度（km）：6
预计游览时间：半天

《中国青年》编辑部旧址→《新青年》编辑部旧址→中华职业教育社旧址→大同幼稚园旧址→《新少年报》社旧址→中国科学社暨明复图书馆旧址

#3 **领略名人风采**

类型：红色文化
线路长度（km）：4.5
预计游览时间：半天

沙千里旧居→杨杏佛旧居→孙中山上海行馆→上海中山故居→中国共产党代表团驻沪办事处（周公馆）旧址→韬奋故居→庆贺鲁迅 50 寿诞集会处旧址

#4 **奏响工运凯歌**

类型：红色文化
线路长度（km）：5.5
预计游览时间：半天

江南制造总局旧址→上海工人第三次武装起义时工人纠察队沪南总部三山会馆→火警钟楼和上海救火联合会旧址→上海工人第三次武装起义发布命令地点

#5 **聆听胜利号角**

类型：红色文化
线路长度（km）：4
预计游览时间：半天

黄浦剧场→老永安公司→五卅运动爱国群众流血牺牲地点→中共“六大”以后党中央政治局机关旧址→南京路上好八连→上海市历史博物馆

#6 **回望会馆风云**

类型：江南文化
线路长度（km）：6
预计游览时间：半天

商船会馆→三山会馆→茶叶公所→梨园公所旧址→四明公所→三山福宁会馆→沪南钱业公所

#7 **品尝舌尖美味**

类型：江南文化
线路长度（km）：3.5
预计游览时间：半天

荣顺菜馆→老城隍庙梨膏糖店→湖心亭→南翔馒头店→绿波廊→松云楼→春风松月楼→上海五香豆商店→宁波汤团店

#8 **讲述城厢故事**

类型：江南文化
线路长度（km）：3.5
预计游览时间：半天

露香园路城墙→上海古城墙和大境道观→上海书店遗址→万竹小学堂（上海实验）→上海梨园公所旧址→上海慈修庵

#9 **寻觅城市原点**

类型：海派文化
线路长度（km）：4.5
预计游览时间：半天

国际饭店→西侨青年会→金门饭店→大新公司→先施公司→老永安公司→新永安公司

#10 **沉浸剧院魅影**

类型：海派文化
线路长度（km）：4
预计游览时间：半天

黄浦剧场→中国大剧院→共舞台→大世界游乐场→南京大戏院（上海音乐厅）

#11 **领略万国风采**

类型：海派文化
线路长度（km）：5
预计游览时间：半天

外白渡桥→上海划船总会旧址→外滩源33（外滩源壹号，英国领事馆）→新天安堂→真光大楼→兰心大楼→女青年会大楼→外滩建筑群→上海电信博物馆

# 徐 汇 区

#12 **自强之路 国歌诞生**

类型：红色文化
线路长度（km）：7.3
预计游览时间：全天

田汉铜像→田汉寓所→马叙伦旧居→聂耳音乐广场→百代公司旧址→上海电影博物馆

#13 **复兴之路 卓越水岸**

类型：红色文化
线路长度（km）：3
预计游览时间：全天

龙华革命纪念地→龙美术馆（西岸馆）→余德耀美术馆→上海香成摄影艺术中心→西岸美术馆

#14 **红色启蒙校园传奇**

类型：红色文化
线路长度（km）：1
预计游览时间：3 小时

南洋公学历史遗迹区→杨大雄烈士纪念碑→史穆烈士墓→五卅纪念柱

#15 **荜路蓝缕砥砺前行**

类型：红色文化
线路长度（km）：7.55
预计游览时间：半天

龙华烈士纪念馆→新四军驻上海办事处旧址→中共江苏省委机关旧址→中共地下党秘密电台旧址→ 66 梧桐苑 · 邻里汇

#16 **梧桐树下的爱国心**

类型：红色文化
线路长度（km）：3
预计游览时间：半天

夏衍故居→张乐平故居→巴金故居→上海宋庆龄故居纪念馆→钱学森图书馆

#17 **科技兴国教育强国**

类型：红色文化、海派文化
线路长度（km）：5.5
预计游览时间：全天

徐光启纪念馆→上海交通大学→钱学森图书馆→中国科学院上海分院→上海音乐学院

#18 **上海女儿 伟大女性**

类型：红色文化
线路长度（km）：5.3
预计游览时间：半天

宋庆龄桃江路旧居→中国福利会旧址→中国福利基金会旧址→上海宋庆龄故居纪念馆

#19 **统一战线 同舟共济**

类型：红色文化
线路长度（km）：5.7
预计游览时间：全天

土山湾博物馆（马相伯纪念地）→汇学博物馆→沈钧儒纪念地→上海宋庆龄故居纪念馆→张澜旧居→马叙伦旧居→统战文化广场

#20 **魅力衡复 旧貌新颜**

类型：红色文化、海派文化
线路长度（km）：2.4
预计游览时间：3 小时

武康大楼→柯灵故居→衡复风貌馆→黑石 M+ 幸福集荟→统战文化广场→夏衍旧居→建业里

#21 **海派文化 之源**

类型：海派文化
线路长度（km）：2
预计游览时间：全天

徐汇公学旧址→徐家汇藏书楼→徐家汇天主教堂→徐家汇观象台→光启公园→土山湾博物馆

#22 **百年人文回眸之路**

类型：红色文化
线路长度（km）：1.3
预计游览时间：半天

衡复风貌馆→柯灵故居→巴金故居→黄兴旧居（老房子艺术中心）→上海宋庆龄故居纪念馆

# 长 宁 区

#23 **红色印迹**

类型：红色文化
线路长度（km）：3.5
预计游览时间：全天

中共中央上海局机关旧址 → 施蛰存旧居 → 愚园路历史名人墙 → 钱学森旧居 → 路易・艾黎旧居 → 长宁区革命文物陈列馆暨《布尔塞维克》编辑部旧址 → 圣约翰

大学交谊楼（解放上海第一宿营地）→ 上海凝聚力工程博物馆

#24 **孙宋的上海故事**

类型：红色文化
线路长度（km）：8
预计游览时间：全天

宋庆龄纪念馆 → 中秋游园会旧址（中央银行俱乐部旧址）→ 路易・艾黎旧居 → 圣约翰大学思颜堂 → 孙科住宅

#25 **悦・读长宁**

类型：海派文化
线路长度（km）：15
预计游览时间：半天至一天

长宁区少年儿童图书馆（东馆）→ 幸福集荟 → 中版书房 → 长宁区图书馆 → 百新书局（缤谷广场店）→ OmS 戏剧图书馆 → 古北市民中心天空书苑 → 长宁区少年儿童图书馆（西馆）

#26 **艺海拾珠**

类型：海派文化
线路长度（km）：8
预计游览时间：半天至一天

上海东方陶瓷美术馆 → 上海杨培明宣传画收藏艺术馆 → 刘海粟美术馆 → 上海艺术品博物馆 → 程十发美术馆 → 上海油画雕塑院美术馆 → 上海国际舞蹈中心 → 虹桥当代艺术馆（长宁文化艺术中心）

#27 **艺术愚园**

类型：海派文化
线路长度（km）：2
预计游览时间：半天至一天

默空间 → 愚园百货公司 → 粟上海 → 愚园公共市集 → 故事门市部 → 长宁区少年宫 → 飞哟画廊 → 愚巷

#28 **人文新华**

类型：海派文化
线路长度（km）：4.5
预计游览时间：半天至一天

红庄 → 上海影城 → 新华别墅（外国弄堂）→ 上生·新所（孙科住宅、哥伦比亚乡村俱乐部）→ 上海民族乐团（陈纳德、陈香梅、董竹君旧居）→ 幸福里

#29 **静雅武夷**

类型：海派文化
线路长度（km）：4
预计游览时间：半天至一天

汤山村 → 武夷村 → 有味三余艺术人文书吧 → 新海艺廊 → 鹿园海派砚雕展示馆 → 丝享荟 → 仁义新村 → 城市会客厅 → 孝义新村

#30 **博物馆之声**

类型：海派文化
线路长度（km）：10
预计游览时间：半天至一天

打字机博物馆 → 上海消防博物馆 → 宋庆龄纪念馆 → 上海儿童博物馆 → 上海纺织服饰博物馆 → 羽瓦台美术馆

#31 **苏河寻音**

类型：海派文化
线路长度（km）：11
预计游览时间：半天至一天

育音堂音乐公园 → 中山公园肖邦雕像 → 苏州河景观步廊长宁段 → 弹指之间 TZ House → 临空音乐公园

#32 **非遗之旅**

类型：江南文化
线路长度（km）：15
预计游览时间：半天至一天
长宁民俗文化中心 → 鹿园海派砚雕展示馆 → 瀚艺海派服饰陈列馆 → 曲全立非遗影像公共体验基地

# 静 安 区

#33 **初心不忘 走向未来**

类型：红色文化
线路长度（km）：3
预计游览时间：2.5 小时

地铁 12 号线南京西路茂名路口→丰盛里—上海毛泽东旧居陈列馆（上海茂名路毛主席旧居，入内）→中共二大会址纪念馆(中国共产党第二次全国代表大会旧址，入内)→延中绿地（中国社会主义青年团中央机关遗址（外观）→八路军驻沪办事处（兼新四军驻沪办事处)旧址（外观））→查公馆（外观）

#34 **璀璨荣光 辉煌巨变**

类型：红色文化
线路长度（km）：3.6
预计游览时间：2.5

上海铁路博物馆（入内)→（路径：山西路→山西大戏院→吴昌硕故居→河南路，沿途介绍)→上海总商会旧址(外观)→(路径：河南路桥→福建路桥→浙江路桥→远眺中共三大后中央局机关历史纪念馆)→上海中国银行办事所及堆栈旧址（外观）→上海四行仓库抗战纪念馆（四行仓库抗战旧址，外观）→乌镇路桥一新闸路桥（远眺福新面粉→厂及堆栈旧址）→中国劳动组合书记部旧址陈列馆（中国劳动组合书记部旧址，入内）

#35 **风云百年 丰碑闪耀**

类型：海派文化
线路长度（km）：1.6
预计游览时间：2 小时

陕西北路中国历史文化名街展示咨询中心→西摩会堂（上海市教委教学研究室）（外观）→崇德女中旧址（同济大学附属七一中学）（外观）→何东公馆（上海辞书出版社）（外观）→怀恩堂（外观）→上海大学旧址（外观）→中华老字号一条街（外观）→报业集团（外观）→静安别墅（外观）→太阳公寓（外观）→民立中学（外观）→上海毛泽东旧居陈列馆（上海茂名路毛主席旧居，入内）

#36 **岁月峥嵘 革命礼赞**

类型：海派文化
线路长度（km）：3.7
预计游览时间：2.5 小时

静安公园→上天桥（360 度眺望）→蔡元培故居（入内）→静安区文化馆（外观）→涌泉坊（外观）→市西中学（外观）→静安寺救火会旧址（外观）→新中国第一家证券营业部旧址→百乐门舞厅（外观）→中共上海地下组织斗争史陈列馆暨刘长胜故居（入内）

#37 **星星之火可以燎原**

类型：红色文化
线路长度（km）：2.5
预计游览时间：2 小时

艺海剧院门口→元利当铺旧址博物馆（入内）→彭湃烈士在沪革命活动地点（外观）→中国劳动组合书记部旧址陈列馆（中国劳动组合书记部旧址）（外观）→中共淞浦特委机关旧址陈列馆（中共淞浦特委办公地点旧址）（入内）→中共中央组织部遗址（外观）→静安雕塑公园

# 虹 口 区

#38 **品・建筑之美**

类型：红色文化→ 海派文化
线路长度（km）：2.3
预计游览时间：半天

李白烈士故居→孔公馆→多伦路 215 号→鸿德堂→长春公寓→内山书店旧址→鲁迅故居→瞿秋白故居→黄竞武烈士故居

#39 **悟・红色经典**

类型：红色文化→ 海派文化
线路长度（km）：3.8
预计游览时间：半天

上海鲁迅纪念馆→鲁迅故居→左联会址纪念馆→中共四大纪念馆→李白烈士故居→ 1933 老场坊

#40 **赏・海派文博**

类型：红色文化→ 海派文化
线路长度（km）：9.5
预计游览时间：全天

鲁迅故居→沈尹默故居→中共四大纪念馆→上海多伦现代美术馆→ 1933 老场坊→上海邮政博物馆→上海犹太难民纪念馆→下海庙

#41 **玩・创意体验**

类型：红色文化→ 海派文化
线路长度（km）：4.3
预计游览时间：半天

1933 老场坊→多伦路文化名人街→花园坊 - 节能环保产业园→空间 188 创意园区→法兰桥创意园区

#42 **乐・都市休闲**

类型：红色文化→ 海派文化
线路长度（km）：3.8
预计游览时间：半天

1933 老场坊→多伦路文化名人街→甜爱

路→上港邮轮城

#43 **访·名人足迹**

类型：红色文化
线路长度（km）：2
预计游览时间：半天

上海鲁迅纪念馆→鲁迅故居→李白烈士故居→沈尹默故居→多伦路文化名人街

#44 **听·虹口往事**

类型：红色文化→ 海派文化
线路长度（km）：4.2
预计游览时间：半天

山阴路历史文化风貌区→多伦路文化名人街→中国证券博物馆

#45 **忆·浦江风云**

类型：红色文化→ 海派文化
线路长度（km）：4.4
预计游览时间：全天

上海犹太难民纪念馆→上港邮轮城→外白渡桥→外滩观光隧道→上海城市历史发展陈列馆→东方明珠广播电视塔零米大厅

#46 **出·埃及游记**

类型：红色文化、海派文化
线路长度（km）：8.2
预计游览时间：全天

和平饭店→河滨大楼→长春公寓→上海展

览中心→市少年宫→马勒别墅→友邦大楼→摩西会堂

#47 **探·力量之源**

类型：红色文化→江南文化→ 海派文化
线路长度（km）：2.3
预计游览时间：半天

鲁迅故居→沈尹默故居→中共四大纪念馆→上海精武体育总会→上海多伦现代美术馆→多伦路文化名人街

#48 **爱·黄金时代**

类型：红色文化→ 海派文化
线路长度（km）：2.3
预计游览时间：半天

孔公馆→左联会址纪念馆→公啡咖啡馆旧址→内山书店→瞿秋白故居→鲁迅故居→鲁迅公园

#49 **绘·都市水乡**

类型：江南文化→ 海派文化
线路长度（km）：1
预计游览时间：半天

石库门建筑群→半岛湾时尚文化创意园→ 1933 老场坊

#50 **游·虹口方舟**

类型：红色文化
线路长度（km）：1
预计游览时间：半天

犹太难民收容所旧址→美犹联合救济委员会旧址→白马咖啡馆→罗伊屋顶花园餐厅→舟山路建筑群→霍山公园→艺术家们的第二故乡（布鲁赫／卫登堡故居）→霍山路第二小学→上海犹太难民纪念馆

### #51 游·海上方舟

类型：红色文化→海派文化

线路长度（km）：2

预计游览时间：半天

上海犹太难民纪念馆→艺术家们的第二故乡（布鲁赫／卫登堡故居）→舟山路建筑群→霍山公园→美犹联合救济委员会旧址→远东反战大会旧址→霍山路第二小学→提篮桥监狱→白马咖啡馆→下海庙→三益村

# 杨 浦 区

#52 **人民城市秀带之旅**

类型：海派文化
线路长度（km）：6
预计游览时间：3 小时

黄浦码头旧址→毛麻仓库旧址→杨树浦水厂栈桥→杨树浦纱厂大班住宅→雨水花园→东方渔人码头→祥泰木行旧址→杨树浦驿站人人屋党群服务站→绿之丘→杨浦大桥→皂梦空间→上海国际时尚中心（原上海第十七棉纺织厂）

#53 **初心启航信仰之旅**

类型：红色文化
线路长度（km）：9
预计游览时间：半天至一天

秦皇岛路驿站→秦皇岛路游船码头（信仰之帆）→浮雕墙→中国救捞陈列馆→王孝和烈士雕像→沪东工人运动史展（信仰之路）→国歌展示馆（信仰之歌）→《共产党宣言》展示馆（信仰之源）→上海院士风采馆（信仰之魂）

#54 **杨浦百年工业之旅**

类型：海派文化
线路长度（km）：5.5

预计游览时间：3 小时

黄埔码头旧址→瑞镕船厂旧址→英商怡和纱厂旧址→杨浦区水厂→上海第一鱼货交易市场→祥泰木行旧址→上海电站辅机西厂旧址→日商上海纺织株式会社旧址→上海电站辅机东厂旧址→中国肥皂公司旧址→上海煤气公司杨树浦工厂旧址→日商大康纱厂旧址→上海工部局电气处新厂旧址→裕丰纺织株式会社旧址

#55 **百年大学书香之旅**

类型：海派文化
线路长度（km）：14
预计游览时间：半天

同济大学（同济大学博物馆）→复旦大学（复旦大学博物馆、复旦大学校史馆、相辉堂）→上海财经大学（毓秀楼、上海财经大学博物馆）→上海体育学院（中国武术博物馆）→上海院士风采馆→上海理工大学（沪江大学历史建筑群）→上海印刷博物馆

#56 **百年市政人文之旅**

类型：海派文化
线路长度（km）：3
预计游览时间：2 小时

绿瓦大楼→长海医院→飞机楼→杨浦区图书馆→江湾体育场

## #57 五角场休闲之旅

类型：其他
线路长度（km）：7
预计游览时间：半天

五角场广场→优迈购物中心（时间车站书店）→合生汇（大众书局）→万达广场（上海书城）→太平洋森活天地（特色小店）→创智天地（大隐书局）→大学路（特色文化商业街）→复旦志达书店→同济书店

## #58 杨浦博物馆之旅

类型：其他
线路长度（km）：22
预计游览时间：半天

国歌展示馆→中国现代国之宝艺术馆→中国烟草博物馆→上海自来水科技馆→中国救捞陈列馆→上海海洋大学博物馆→上海印刷博物馆→上海院士风采馆→中国武术博物馆→复旦大学博物馆→同济大学博物馆

## #59 滨江工业遗存之旅

类型：海派文化
线路长度（km）：5.5
预计游览时间：3 小时

秦皇岛路旅游咨询服务中心→船坞秀场→跑者驿站→水厂栈桥→雨水花园驿站→天外之物→东方渔人码头→自由方块→安浦路桥→观演剧场→城市的野生→人人屋驿站→绿之丘→时间之载→大桥

公园驿站→山→生态水池→共生构架→皂梦空间→ encounter →沙滩排球场→码头篮球场→露天旱冰场→轻舟过隙→纱泉广场→徊→黄浦货仓→电厂驿站→起重机的对角线→上海国际时尚中心→定海路桥

#60 **活力水岸亲子之旅**

类型：其他
线路长度（km）：4.2
预计游览时间：2—3 小时

上海国际时尚中心（珍得巧克力剧院）→杨浦电厂遗迹公园→上棉十二厂（纱泉广场、卡其乐园）→杨浦煤气厂（边园、码头篮球场、沙滩排球场）→上海制皂厂（皂梦空间）→共生构架

#61 **江湾历史风貌之旅**

类型：海派文化
线路长度（km）：9.5
预计游览时间：半天至一天

绿瓦大楼→中国武术博物馆→杨浦区图书馆→江湾体育场→复旦大学校史馆、相辉堂→同济大学博物馆→沪江大学（上海理工大学）

#62 **城市空间艺术之旅**

类型：其他
线路长度（km）：4.2
预计游览时间：2—3 小时

城市空间艺术季展品：天外之物→方块花园→城市的野生→时间之载→山→相遇ENCOUNTER→轻舟过隙→徊→黄浦货舱→起重机的对角线

#63 **娱悦休闲亲子之旅**

类型：其他
线路长度（km）：12
预计游览时间：半天至一天

安徒生童话乐园→上海共青森林公园→珍得巧克力剧院→皂梦空间→家家乐梦幻乐园

# 普 陀 区

#64 **沪西故事**

类型：红色文化
线路长度（km）：5.6
预计游览时间：2—4 小时

长寿公园→沪西革命史陈列馆→中央造币厂旧址（外观）→顾正红纪念馆→上海纺织博物馆、申九“二·二”斗争纪念地

#65 **水岸寻访**

类型：海派文化
线路长度（km）：10
预计游览时间：1 天

苏州河梦清园环保主题公园→宜昌路救火会→苏州河工业文明展示馆→苏宁艺术馆→上海科技金融博物馆→上海印染机械厂旧址

#66 **大隐于市**

类型：江南文化
线路长度（km）：8
预计游览时间：4 小时

玉佛禅寺→上海元代水闸遗址博物馆→曹杨新村村史馆→真如寺→真如羊肉馆

#67 **行走苏州河畔**

类型：海派文化
线路长度（km）：12
预计游览时间：1 天

长寿段：信和纱厂旧址→阜丰福新面粉厂旧址（外观）→上海啤酒有限公司旧址→中央造币厂旧址（外观）→宜昌路救火会大楼旧址→江苏药水厂旧址（外观）→福新第三面粉厂旧址（外观）

#68 **行走苏州河畔**

类型：海派文化
线路长度（km）：3
预计游览时间：2 小时

长风段：上海印染机械厂旧址→苏州河工业文明展示馆→上海试剂总厂烟囱

#69 **行走澳门路**

类型：红色文化、海派文化
线路长度（km）：2
预计游览时间：2 小时

信和纱厂旧址→上海纺织博物馆、申九“二·二”斗争纪念地→顾正红纪念馆→中华 1912 创意园（外观）→澳门路日式住宅（外观）

# 浦东新区

#70 **行走陆家嘴**

类型：红色文化、海派文化
线路长度（km）：2
预计游览时间：1.5 小时

浦东开发陈列馆→陆家嘴→吴昌硕纪念馆

#71 **传承闻天精神之旅**

类型：红色文化
线路长度（km）：1
预计游览时间：1 小时

张闻天故居

#72 **古韵川沙游**

类型：江南文化
线路长度（km）：1
预计游览时间：2 小时

中市街牌坊→川沙戏曲艺术展示中心→川沙营造馆→川沙古城墙，黄炎培故居

# 宝 山 区

#73 **筑梦之旅**

类型：红色文化
线路长度（km）：7.5
预计游览时间：1 天

上海淞沪抗战纪念馆→上海解放纪念馆→吴淞口国际邮轮港→上海长江河口科技馆→宝山城市规划展示馆

#74 **振兴之旅**

类型：海派文化
线路长度（km）：12
预计游览时间：1 天

半岛 1919 文创园（大中华纱厂及华丰纱

厂旧址)→中成智谷→上海玻璃博物馆→智慧湾科创园—中国 3D 打印文化博物馆

#75 **文复之旅**

类型：江南文化
线路长度（km）：13.9
预计游览时间：1 天

上海市陶行知纪念馆→上海木文化博物馆→上海宝山国际民间艺术博览馆→龙现代艺术中心

# 闵 行 区

#76 **七宝老街线**

类型：江南文化→ 海派文化
线路长度（km）：1.5
预计游览时间：半天

蒲汇塘桥→七一人民公社旧址→解元厅→七宝茶馆书场

#77 **召稼楼线**

类型：江南文化→ 海派文化
线路长度（km）：3

预计游览时间：3 小时

梅园→奚氏宁俭堂宅院→赵元昌商号宅院→奚家恭寿堂住宅→礼耕堂→道南桥→奚世瑜住宅

#78 **闵行文化公园线**

类型：红色文化、江南文化
海派文化和其他
线路长度（km）：2.3
预计游览时间：半天

漕宝路七号桥碉堡→闵行博物馆→宝龙美术馆

#79 **马桥线**

类型：红色文化、海派文化
线路长度（km）：1
预计游览时间：2 小时

马桥文化展示馆和俞塘民众教育纪念馆

#80 **江川路街道线**

类型：江南文化、海派文化
线路长度（km）：15
预计游览时间：半天

闵行一条街→项家宅院→古藤园

# 嘉 定 区

#81 **疁城古韵之旅**

类型：江南文化
线路长度（km）：2
预计游览时间：2 小时

嘉定孔庙→法华塔→嘉定别墅（竹刻博物馆）

#82 **嘉定新城时尚之旅**

类型：海派文化
线路长度（km）：2
预计游览时间：2 小时

上海保利大剧院→嘉定图书馆→嘉定文化馆

#83 **汽车城之旅**

类型：其他
线路长度（km）：2
预计游览时间：2 小时

汽车博物馆、汽车会展中心

#84 **F1 赛车场之旅**

类型：其他
线路长度（km）：1
预计游览时间：1 小时

上海国际赛车场

#85 **江南园林之旅**

类型：江南文化
线路长度（km）：2
预计游览时间：2 小时

汇龙潭、秋霞圃公园及建筑

# 金 山 区

#86 **枫泾红色记忆之旅**

类型：红色文化
线路长度（km）：4
预计游览时间：3 小时

袁世钊故居→陆龙飞烈士墓→朱学范墓→朱学范故居→消防纪念塔→东区火政会→侵华日军飞机枪弹遗迹墙→人民公社旧址

#87 **枫泾江南文化之旅**

类型：江南文化
线路长度（km）：4
预计游览时间：3 小时

致和桥→宝源桥→秀兴桥→来源桥→跻云桥→庆云桥→瑞虹桥→程十发祖居→惠安

桥→枫泾镇邮电局旧址

#88 **金山南部红色记忆之旅**

类型：红色文化
线路长度（km）：9
预计游览时间：3 小时

金山卫城南门侵华日军登陆点→金山区烈士陵园→新街暴动纪念碑→高天梅故居遗址→姚光故居

# 青 浦 区

#89 **桥乡金泽**

类型：江南文化
线路长度（km）：2
预计游览时间：2 小时

普济桥→林老桥→金泽放生桥→迎祥桥→如意桥→颐浩寺遗址→万安桥→天皇阁桥→金泽陈家仓库

#90 **红色练塘**

类型：红色文化、江南文化
线路长度（km）：2

预计游览时间：3 小时

永兴桥→义学桥→朝真桥→顺德桥→陈云纪念馆→颜安小学老教室→杜衡伯纪念塔→吴开先旧居→阜康酱园→圣堂→东区救火会→吴志喜故居→练塘镇下塘街 25 号民居→练塘下塘街街廊

#91 **诗画朱家角**

类型：江南文化
线路长度（km）：4.5
预计游览时间：4 小时

放生桥→泰安桥→福星桥→中和桥→永丰桥→永安桥→课植园→涵大隆酱园→大清邮局旧址→童天和国药号→西湖街金宅→东井茶楼→东井街吴氏住宅→朱家角俱乐部茶楼→朱家角基督堂→西井街舒宅→柳亚子别墅→朱家角城隍庙→圆津禅院

# 松 江 区

## #92 文化寻根之旅

类型：海派文化
线路长度（km）：3.5
预计游览时间：1 天

佘山天主教堂→佘山天文台→广富林文化遗址

## #93 云间邦彦之旅

类型：红色文化
线路长度（km）：15
预计游览时间：1 天

马相伯故居→陈化成祠→兰瑞堂→陈子龙墓→夏允彝夏完淳父子墓

## #94 园林盘桓之旅

类型：其他
线路长度（km）：4
预计游览时间：1 天

颐园→醉白池→方塔园

## #95 塔刹探奇之旅

类型：其他
线路长度（km）：15
预计游览时间：1 天

秀道者塔→李塔→唐经幢→兴圣教寺塔

#96 **建筑览胜之旅**

类型：江南文化
线路长度（km）：6
预计游览时间：1 天

大仓桥→杜氏雕花→方塔园→云间第一楼

# 奉 贤 区

#97 **亲子滨海 2 日游**

类型：海派文化
线路长度（km）：20
预计游览时间：2 天

华亭海塘→东海观音寺

#98 **红色文化之旅**

类型：红色文化
线路长度（km）：20
预计游览时间：1 天

上海中国人民志愿军纪念馆→李主一烈士纪念碑→中共奉贤县委旧址→上海海湾国家森林公园

#99 **人文艺术之旅**

类型：海派文化→红色文化→江南文化
线路长度（km）：5
预计游览时间：1 天

沈家花园→卜罗德祠→南塘第一桥→奉贤博物馆

# 崇 明 区

#100 **古韵悠悠 东海瀛洲**

类型：红色文化、江南文化
线路长度（km）：5
预计游览时间：1 天

崇明岛岛碑→崇明区博物馆→瀛洲公园→金鳌山

#101 **绿色生态 人文崇明**

类型：海岛文化
线路长度（km）：3
预计游览时间：半天

崇明区图书馆→崇明生态科技馆→崇明美术馆→崇明区规划展示馆

别册设计：徐 翔

扫码打开

这 里 是 上 海

建筑可阅读 有声书

上海广播电视台

“侧耳”诵读